ORGANELLES IN TUMOR DIAGNOSIS: AN ULTRASTRUCTURAL ATLAS

ORGANELLES IN TUMOR DIAGNOSIS: AN ULTRASTRUCTURAL ATLAS

Brian Eyden B.Sc. Ph.D.

Clinical Scientist and Head of Electron Microscopy
Department of Histopathology
Christie Hospital National Health Service Trust
Manchester, England

IGAKU-SHOIN NEW YORK * TOKYO

Published and distributed by

IGAKU-SHOIN Medical Publishers, Inc.,
One Madison Avenue, New York, New York 10010

IGAKU-SHOIN Ltd.,
5-24-3 Hongo, Bunkyo-ku, Tokyo 113-91.

Library of Congress Cataloging-in-Publication Data

Eyden, Brian.
Organelles in tumor diagnosis: an ultrastructural atlas/Brian Eyden.
p. cm.
Includes bibliographical references and index.
1. Tumors—Cytodiagnosis—Atlases. 2. Diagnosis, Electron microscopic—Atlases. 3. Tumors—Ultrastructure—Atlases. 4. Cell organelles—Atlases. 5. Cancer cells—Atlases. I. Title.
[DNLM: 1. Organelles—ultrastructure—atlases. 2. Neoplasms—diagnosis—atlases. OH 591 E97a 1996]
RC270.3.E43E96 1996
616.99′207582—dc20
DNLM/DLC
for Library of Congress

96-1857
CIP

ISBN: 0-89640-311-4 (New York)
ISBN: 4-260-14311-5 (Tokyo)

Printed and bound in the U.S.A.
10 9 8 7 6 5 4 3 2 1

Acknowledgments

I would like to express my sincere thanks to the many colleagues who have helped me with this book. Thanks, first of all, go to my consultant pathologist colleagues at the Christie Hospital, Dr. Sankar Banerjee and Dr. Martin Harris, whose expertise in classical pathology has given an overall perspective for my own purely ultrastructural interpretations. I thank the Histology staff at the Christie Hospital for attending to the technical and organizational details of specimen referral, and I thank my undergraduate students for carrying out the routine technical work of the Unit. Among the latter, I particularly thank Mrs. Daxa Jobanputra for taking the cover photograph of the Langerhans cell granules.

To these I add the many friends in the Society of Ultrastructural Pathology with whom over the years I have enjoyed amicable and instructive discussions—especially Irving Dardick of Toronto, who has always been a source of encouragement. To the colleagues and friends who have been generous enough to give their time to read chapters of this book and contribute constructive criticism—Per Carstens, Irving Dardick, Karlene Hewan-Lowe, and Richelle and Douglas Weeks—as well as the colleagues who have generously allowed me to use originally published micrographs or who have entrusted me with resin blocks, I offer my sincere gratitude.

The staff of Igaku-Shoin in New York deserve my sincere thanks for their competence, courtesy, and patience. Last but far from least, I thank my family—Freda, Joanna, and Suzy—who have supported and tolerated me during the preparation of this work.

BRIAN EYDEN

Preface

Electron microscopy and immunocytochemistry are the two most important techniques ancillary to conventional histology and traditional special stains for solving diagnostic problems in human tumors. Electron microscopy has been diagnostically useful since the 1970s. Immunocytochemistry is a more recent development and is regarded by many pathologists as the more convenient and important. It is now clear, however, that the antibody specificity—on which the technique is based and which was considered so useful in the early days of immunocytochemistry—has not stood the test of time. Tumor immunophenotype is currently recognized as far more complex than initially thought, and one relatively new area where electron microscopy is useful is to clarify the nature of tumors showing unexpected or complex immunostaining patterns. Many now feel that the combined use of electron microscopy and immunocytochemistry offers the best approach to understanding tumor cell differentiation and achieving maximum diagnostic accuracy.

Electron microscopy in human tumor diagnosis is rarely carried out separately from other diagnostic disciplines—particularly, clinical assessment, histology, and immunocytochemistry; and the successful application of electron microscopy, while making use of information from these sources, at the same time requires the following special areas of knowledge: (1) being able to identify images in micrographs as distinctive types of cell organelle with an appropriate nomenclature; (2) having a knowledge of the distribution of these organelles in specific types of cell, tissues, and lesion; (3) having a knowledge of the overall ultrastructural characteristics of named tumors or lesions (this is only possible with a basic understanding of 1 and 2).

This book is concerned primarily with the first of these areas. The recently published *Handbook of Diagnostic Electron Microscopy for Pathologists-in-Training*, published by Igaku-Shoin, provides a first teaching text for student pathologists. In addition, several monographs are available on the fine structural features of named tumors and lesions. The present atlas is designed to bridge the gap between these two areas.

The main purpose of this book is to provide descriptions of the cell components seen in human tumors to enable ultrastructural pathologists to identify them as organelles with an appropriate name and significance. With regard to this latter point, information is also given on cell organelle distribution and specificity.* Most importantly, this atlas deals with analyzing images from the field of *practical routine* diagnostic

* A simple table is provided in Chapter 1 listing the main groups of tumors and their characteristic organelles; while in the chapters dealing with specific organelles, text devoted to information on organelle distribution and specificity is highlighted by being indented and being set in smaller type. However, the literature in this area itself is so extensive that it may well require a quite separate and subsequent volume.

work. It therefore not only illustrates the typical images from nicely fixed tissues that are the norm in other currently available texts, but also pays attention to the *atypical* images shown by tumor organelles—these, after all, are the ones more likely to cause interpretational difficulty. Such atypical appearances may reflect inherent biological abnormality or may result from poor preservation.

A further objective is to emphasize suitable terminology. A multiplicity of terms is encountered in the literature for many of the organelles observed in tumors, and not all of them are appropriate. Effective communication between pathologists can only be enhanced with a uniform nomenclature, and I hope that this atlas will help to promote this at the level of cell structure.

It is important to emphasize also that not all organelles observed in tumor specimens are diagnostically significant. Generally, one sees organelles in two main contexts: mostly, one *searches* for a specific organelle to confirm a suspected diagnosis, but one also *encounters* organelles which are not the object of a specific search. In any given instance, these organelles may or may not be diagnostically important; in order to make this assessment, however, it is desirable to have as wide a knowledge as possible of human cell ultrastructure.

I have cited some of the published literature as a means of illustrating examples of organelle ultrastructure where space has not been available to show them in the atlas itself. The citations are intentionally selective and have been kept to a minimum. Some particular points of discussion in the text were felt to require individual referencing. All of these citations are listed in the references section.

Not all of these references—and in fact not all of the illustrations—are of neoplastic cells: some are from normal or reactive cells. To an extent, organelle interpretation in tumors is based on normal cell structure; one confidently identifies melanosomes in malignant melanoma or desmosomes in squamous cell carcinoma, for example, by noting the similarity of tumor melanosomes and desmosomes to those found in normal melanocytes and keratinocytes, respectively. At the same time, however, one recognizes the existence of structurally abnormal organelles in tumor cells. Therefore, the information contained in the atlas forms a basis for interpreting ultrastructure on a wider basis than just neoplasia, but the emphasis of the book is on assisting in the interpretation of tumor ultrastructure.

How this atlas is to be used will depend on the ultrastructural experience of the pathologist. For those in training to be histopathologists, the atlas is intended to provide appropriate starting guidelines in organelle recognition and terminology; reading the book will help to develop a detailed knowledge of the appearances of organelles likely to be seen or needed in tumor diagnosis. The simplified table in Chapter 1 may also be useful to alert pathology residents to the organelles characteristic of particular tumor groups. By contrast, for experienced ultrastructural pathologists this atlas may provide an opportunity for refining or updating their current knowledge and interpretations. The overall objective is to provide a handy and practical volume to help pathologists involved in day-to-day routine work so that they can give more precise descriptions and interpretations, and consequently better diagnoses—ultimately, one hopes, to the benefit of patient management.

BRIAN EYDEN

Contents

1

Introductory Remarks on Diagnosing Tumors by Electron Microscopy

GLOBAL FACTORS INFLUENCING INTERPRETATION

Using electron microscopy in tumor diagnosis requires an ability to recognize cell structures as markers for a given type of cell, tissue, or lesion. For example, the elliptical bodies with an internal cross-striated lattice (Chapter 10) are identified as melanosomes and are recognized as defining melanocytic differentiation. Often, their inherent features are clearly defined enough to allow unambiguous identification. Some, however, show atypical ultrastructure, either because of the neoplastic process or for technical reasons. In addition, functionally different organelles may show overlapping features: For example, serous granules, exocrine granules of mammary epithelium, neuroendocrine granules, and lysosomes can all assume comparable appearances; similarly, vimentin, desmin, glial, and neuronal intermediate filaments are difficult, if not impossible, to distinguish in purely ultrastructural terms—in short, atypical and overlapping ultrastructure can compromise interpretation. In these situations, the investigator will inevitably look for data from other disciplines to help consolidate a particular interpretational direction. Just as clinical findings can influence a final histological interpretation, it has been found that collective data from clinical, histological, and immunocytochemical disciplines can influence ultrastructural interpretation. An example is provided in Plate 23, where organelles have an appearance of melanosomes but lack a clearly defined internal lattice. These are justifiably interpreted as melanosomes because (1) despite lacking certain aspects of fine structural detail, they possess an overall similarity to melanosomes, and (2) they were found in a clinically, histologically, and immunocytochemically unambiguous malignant melanoma.

Further illustrative examples of where thc ultrastructural interpretation of organelles can be influenced by a collective body of light microscope data include:

Mucigen granules	Carbohydrate stains
Serous/zymogen granules	Cytochemistry and immunostaining for digestive enzymes (e.g., trypsin, chymotrypsin)
Neuroendocrine granules	Argyrophilia and immunostaining for chromogranin and peptide and nonpeptide hormones
Lysosomes	Acid phosphatase cytochemistry
Weibel–Palade bodies	Immunostaining for endothelial cell markers (e.g., factor VIII-related antigen, CD31, CD34)
Mammary exocrine granules	Immunostaining for casein and lactalbumin
Surfactant lamellar bodies	Immunostaining for surfactant apoprotein
Cytoplasmic filaments	Immunostaining for smooth and striated muscle actins, myosins, and intermediate filament proteins
Lamina and fibronexus	Immunostaining for laminin, type IV collagen, and fibronectin
Desmosomes	Immunostaining for desmosome-associated proteins (e.g., desmoplakins, cytokeratins).

It should also be remembered also that there are useful *ultrastructural* stains for demonstrating certain organelles. These can be technically demanding, however, and tend not to be popular in routine tumor diagnosis. The more commonly used ones include (1) dopa reaction and Warthin–Starry technique for melanosomes and melanin and (2) the Grimelius and uranaffin techniques for neuroendocrine granules. In addition, ultrastructural peroxidase cytochemistry has a role in differentiating myeloid and lymphoid proliferations. Finally, it may be proven that the most unambiguous data will be obtained by immunoelectron microscopy and the newly emerging molecular biology techniques (see the symposia in *Ultrastructural Pathology*, volume 16, numbers 1–2, 1992 and in *Human Pathology*, volume 25, number 6, 1994).

ORGANELLE CHARACTERISTICS OF NAMED TUMORS

Table 1 is a simplified listing of the main diagnostically useful organelles characterizing some of the better-known tumors. It is intended to be introductory to ultrastructural tumor diagnosis rather than exhaustive. It points out those chapters in the Atlas providing detailed information on the organelles in question.

Table 1. Main Diagnostically Useful Organelles Characterizing Some of the Better-Known Tumors

Tumor	Organelle	Chapter
Lymphoid and Myeloid Lesions		
Mycosis fungoides/Sézary syndrome	Cerebriform nuclei	2
Lymphoplasmacytoid lymphoma[a]	rER[b]	3
True histiocytic lymphoma	Lysosomes	6, 17
	Multivesicular bodies	15
Plasmacytoma/myeloma	rER	3
	Russell bodies	3
Langerhans cell granulomatosis	Langerhans cell granules	14
Myeloid, monocytic, and monoblastic leukemias; granulocytic sarcoma	Lysosomes	6, 17

continued

Table 1. Main Diagnostically Useful Organelles Characterizing Some of the Better-Known Tumors

Tumor	Organelle	Chapter
Melanocytic Lesions		
Malignant melanoma	Melanosomes	10
Epithelial Tumors		
Basal/squamous carcinoma	Desmosomes	26
	Tonofibrils	24
	Lamina	27
Adenocarcinoma	Desmosomes	26
	Junctional complexes	26
	Lumina	28
	Short microvilli	28
	Lamina	27
	Mucigen granules	7
Alveolar cell carcinoma	Surfactant bodies[c]	19
Acinic cell carcinoma	Serous granules[c]	8
Renal cell carcinoma/clear cell carcinomas	Lipid[c]	19
	Glycogen	20
Sebaceous carcinoma	Lipid[c]	19
Adenoid cystic carcinoma	Stromal spaces with lamina[c]	27
Mesothelioma	Desmosomes	26
	Junctional complexes	26
	Lumina	28
	Long slender microvilli	28
	Lamina	27
Wilms' tumor	Desmosomes	26
	Junctional complexes	26
	Lumina	28
	Lamina	27
Oncocytic tumors	Mitochondria[c]	18
Neuroendocrine carcinoma/ Carcinoid	Neuroendocrine granules[c]	5
Steroidogenic Tumors		
Adrenocortical carcinoma/	Tubular cristae	18
Granulosa cell tumor	Lipid	19
	sER[d]	3
Leydig cell tumor	Crystals[e]	29
Neuroendocrine/Neuronal/CNS Tumors		
Merkel cell tumor	Neuroendocrine granules	5
	Paranuclear intermediate filament aggregates	24
Insulinoma	Neuroendocrine granules with crystalline cores	5
Pheochromocytoma	Eccentric-core granules	5
Neuroblastoma	Neuroendocrine granules	5
	Microtubules in processes	25
Primitive peripheral	Microtubules in processes	25

continued

Table 1. Main Diagnostically Useful Organelles Characterizing Some of the Better-Known Tumors

Tumor	Organelle	Chapter
Neuroectodermal tumor	Neuroendocrine granules	5
Gastrointestinal autonomic nerve tumor	Neuroendocrine granules	5
	Microtubules in processes	25
	Intermediate filaments	24
Ependymoma	Processes, lumina	28
	Cilia	25
Astrocytoma	Processes containing glial intermediate filaments	24
	Rosenthal fibers	24
Oligodendroglioma	Microtubules	25
	Lysosomes	17
Central neurocytoma	Microtubules in processes	25
	Synapses	28
Meningioma	Processes	28
	Desmosomes	26
Mesenchymal Tumors		
Fibroblastic tumors (e.g., fibrosarcoma, malignant fibrous histiocytoma)	rER	3
Myofibroblastic lesions (e.g., nodular fasciitis)	rER	3
	Myofilaments with focal densities	22
	Fibronexus	27
Chondroma/-sarcoma	Fine surface processes	28
	rER	3
	Matrix proteoglycans	30
Extraskeletal myxoid chondrosarcoma	microtubular inclusions in rER	3
Osteoma/-sarcoma	rER	3
Leiomyoma/-sarcoma	Myofilaments with focal densities	22
	Submembranous densities	22
	plasmalemmal caveolae	13
	Lamina	27
Myoepithelioma	Myofilaments with focal densities	22
	Lamina	27
	Desmosomes	26
	Tonofibrils	24
Rhabdomyoma/-sarcoma	Thick and thin myofilaments with Z disks	22
	Submembranous densities	22
	Lamina	27

continued

Table 1. Main Diagnostically Useful Organelles Characterizing Some of the Better-Known Tumors

Tumor	Organelle	Chapter
Schwannoma/malignant schwannoma	Processes	28
	Lamina	27
	Luse bodies	30
Granular cell tumor	Secondary lysosomes	17
Hemangioma/angiosarcoma	Weibel–Palade bodies	11
	Tight junctions	26
	Pinocytotic vesicles	13
	Lumina	28
	Lamina	27
Lipoma/-sarcoma	Lipid	19
	Lamina	27
Miscellaneous Tumors		
Malignant rhabdoid tumor	Intermediate filaments	24
Ewing's tumor	Glycogen	20
Endodermal sinus tumor	Glycogen	20
	Processes, lumina	28
	Amorphous "lamina"	27
Seminoma	Euchromatinic nucleus	2
	Nucleolonematous nucleolus	2
Alveolar soft-part sarcoma	Crystals	29
Epithelioid sarcoma	Intermediate filaments	24

[a] For the nomenclature of lymphoproliferative lesions, the revised European American lymphoma classification of Chan et al. (1995) is used.
[b] rER, rough endoplasmic reticulum.
[c] In addition to general features of carcinoma/adenocarcinoma.
[d] sER, smooth endoplasmic reticulum.
[e] In addition to tubular cristae, sER, and lipid.

TECHNIQUE

All micrographs are of specimens processed according to conventional transmission electron microscope technique. Most have been retrieved from the surgical cut-up in formalin, and fixation was considered complete on arrival of the specimen in the Electron Microscope Laboratory.

Glutaraldehyde fixation was for several hours in 2.5% glutaraldehyde in 0.1 M (pH 7.4) cacodylate buffer, followed by rinsing in buffer. Some tissues were fixed overnight or over the weekend. Specimens were given a secondary fixation in 1% osmium tetroxide in buffer, followed by a further buffer rinse, a brief rinse in 0.1 M sodium acetate, then left in 0.5% aqueous uranyl acetate overnight. Dehydration in graded ethanols and propylene oxide was followed by gradual infiltration with 33% and 67% epoxy resin–propylene oxide mixtures, followed by neat resin. Polymerization was at 60°C for 16 hr. The resin embedding medium was based on the Epon-type monomer Agar 100 from Agar Scientific (England) and was made up as follows:

Agar-100	38.4 g
Dodecenylsuccinic anhydride	20.8 g
Methylnadic anhydride	20.8 g
2,4,6-Tri(dimethylaminomethyl) phenol (DMP 30)	35 drops

Tissue retrieved from paraffin blocks was deparaffinized in xylene, rehydrated to buffer, and then osmicated and processed as above.

Blocks for ultramicrotomy were selected from 1-μm resin sections stained in 1% toluidine blue in 1% aqueous borax solution. Ultrathin sections were cut on a diamond knife, collected on uncoated copper grids and stained in 5% aqueous uranyl acetate (20 min) and then in Reynolds' lead citrate (5 min). They were examined in an AEI electron microscope at 80 kV fitted with a six-grid specimen holder.

MEASUREMENT AND UNITS

Measurements were made using two calibration standards. A multigrid holder (six grids) had the advantage of allowing the operator to photograph a specimen and then allowing him or her to move to a calibration grid without altering any of the conditions affecting magnification. For low and medium powers, a cross-grating replica with 2160 lines per millimeter (spacing, 463 nm) was used; for higher magnifications, a negatively stained catalase crystal grid with periodicities of 6.85 and 8.75 nm was used.

Note the following:

1. Micrometer (μm) and nanometer (nm) are the only permissible units:
   1 mm = 1000 μm and 1 μm = 1000 nm
   Micron (μ), millimicrometer (=millimu = mμ), and angstrom (Å) are obsolete:
   1 μm = 1 μ, 1mμ = 1 nm, 1 nm = 10 Å
2. Since the limit of resolution in the sections typically used in diagnostic transmission electron microscopy is about 2 nm, it is appropriate when recording measurements to round values to the nearest nanometer.

2

Nucleus

The **nucleus** encloses most of the cell's genetic material, its ribosome-synthesizing machinery, and the environment for gene transcription and replication. It is usually the most conspicuous single inclusion in a cell; it therefore catches the eye in histological and resin sections observed under low magnification and provides an instant assessment of tissue cellularity.

Because of the large size of the nucleus and because much of its structure is observable by light microscopy, nuclear ultrastructure is not as diagnostically important as certain other cell constituents such as secretory granules, cytoplasmic filaments, and surface features. Some characteristic nuclear morphology can be usefully demonstrated by electron microscopy—for example, the cerebriform nuclei of mycosis fungoides and of Sézary cells, as well as the labyrinthine nuclei of dermatofibrosarcoma protuberans.

Demarcation of the nucleus from the cytoplasm is due to the overall density produced by the nuclear envelope membranes themselves in combination with associated structures (the nuclear fibrous lamina and the dense granular material contacting the inner surface of the envelope referred to as **heterochromatin**). For clearly defined nuclear profiles, see Plates 1, 10, 11, 13, 14, 16, 17, 20–22, 26, 27, 31, 33, 35, 36, 39, 40, 42, 44, 45, 47–51, 54–57 and 60.

CHROMATIN

Nuclear deoxyribonucleic acid is complexed with protein as **deoxyribonucleoprotein**, which is more or less equivalent to the **chromatin** of light microscopy. Despite the fact that chromatin is a light microscopy term, it is convenient to use it in electron microscopy for deoxyribonucleoprotein. Chromatin has different appearances depending on whether it is being transcribed into messenger RNA or not. In its nontranscribing state, chromatin is believed to be located within granular material referred to as **heterochromatin**. This is often located on the underside of the nuclear envelope (Plate 1A; see also Plates 11A, 11B, 17A, 31A, 34A, 39A, 39B and 50C); therefore the peripheral heterchromatin corresponds with the Feulgen-positive DNA-containing material of light microscopy. The terms heterochromatin, chromatin, and condensed chromatin are often used interchangeably.

Abundant heterochromatin is typically found in:

- Resting (inactive) lymphocytes (e.g., in small lymphocytic lymphomas)
- Small centrocytes (e.g., in follicle center lymphomas)
- Polyclonal plasma cells (in the form of roughly triangular blocks)
- Mast cells, eosinophils, basophils and neutrophils

For examples, see Plates 13A, 14B, and 17A.

Heterochromatin is also found as discrete blocks away from the nuclear envelope. Some are as large as the pieces of envelope-associated heterochromatin (Plates 1A, 11B, 17A, 31A), whereas in other nuclei the heterochromatin can be in the form of many much smaller pieces (Plates 4C, 42A, 45A, and 47A); this gives the nuclear interior a speckled or finely textured appearance. For high-magnification views of the granular texture of heterochromatin, see Plates 39B and 60A.

Transcribing chromatin is referred to as **euchromatin**. This term is frequently used for the comparatively pale-staining part of the nucleus; the latter, however, merely contains euchromatin which is dispersed throughout a nuclear sap or matrix. Therefore, the pale-staining nuclear interior is best designated as **interchromatin**. Pale-staining nuclei are often referred to as **euchromatic** or **euchromatinic** and have virtually no aggregated heterochromatin (Plates 1C, 2A, 3C, and 60B).

The euchromatinic nucleus indicates a cell active in the early steps of protein elaboration and is found in examples of all of the major tumor classes, but the following typically have such nuclei:

- Large-cell lymphoid proliferations (e.g., diffuse large B-cell lymphomas, anaplastic large-cell lymphomas, Hodgkin's disease)
- Seminomas

Pale-staining nuclei which nevertheless lack a typical euchromatinic appearance have been described as **fibrillar** nuclei because they contain a mass of 25-nm fibrils. This organization is seen predominantly in formalin-fixed and postmortem material and may therefore be artifactual (see Yunis et al., 1984).

Some heterochromatin is almost certainly a manipulative or physiological artifact. Tumors with a monomorphic cell population by light microscopy can contain areas where the nuclei are euchromatinic, and they can contain other areas where nuclei have prominent heterochromatin (Plate 1B and 1C); the latter may also have extremely clear interchromatin, suggesting leaching out of biological materials and overall poor preservation. This effect may result from autolysis due to local anoxia or—in the case of formalin fixation—to different sampling sites. Tissue sampled from the surface of a large block fixed in formalin will probably be better preserved than tissue sampled from the interior, in line with the time required for the fixative to reach these internal regions; during this time, autolytic changes are likely. For this reason, it is appropriate when choosing blocks for ultramicrotomy from toluidine blue resin sections to assess nuclear staining with care. Pale-staining inconspicuous nuclei usually turn out to be well-preserved; sharply delineated nuclei are often poorly preserved, because chromatin is artifactually condensed onto the nuclear envelope and soluble materials are leached out of the nuclear interior.

NECROSIS AND APOPTOSIS

Apoptotic cells display distinctive crescentic profiles of very homogeneous peripheral heterochromatin (Plate 1A): a mere abundance of heterochromatin is not enough to indicate apoptosis. Significantly also in apoptotic cells, cytoplasmic preservation may be very good. Apoptotic cells are frequently ingested by epithelial cells or macrophages, and then they assume the appearances of necrotic cells as they become degraded within secondary lysosomes.

Necrotic nuclei (Plate 1D) frequently contain many blocks of heterochromatin, the nuclear envelope may fragment, and the cytoplasm also shows comparable signs of poor preservation [i.e., washed out cytoplasmic matrix (Plate 1D)].

Plate 1. **A:** Two nuclei with differing chromatin patterns. Above, note the presence of a tumor cell with a small but distinct amount of peripheral heterochromatin (*large arrows*), as well as heterochromatin in the form of discrete blocks away from the envelope (*small arrows*). Below, an apoptotic cell has characteristic crescentic profiles of highly condensed and marginal heterochromatin (*arrowheads*). Reactive follicular hyperplasia, cervical node. × 10,300. **B** and **C**: Two regions from the same tumor, which consisted of uniform sheets of spindled cells by light microscopy. In **B**, nuclei show condensed heterochromatin (*arrowheads*) and washed out spaces; in **C**, the nuclei look well-preserved and euchromatinic. The clear cytoplasmic spaces are leached out glycogen. Putative spindle/oval-cell extraskeletal Ewing's tumor, soft tissues of scapular region. Both × 6500. **D**: Necrotic cells showing many discrete blocks of heterochromatin, poorly preserved nuclear envelope, and washed out nuclear interior and cytoplasm (*). Oat cell carcinoma, metastatic to larynx. × 5500. hc, heterochromatin; mi, mitochondria; Nu, nucleolus.

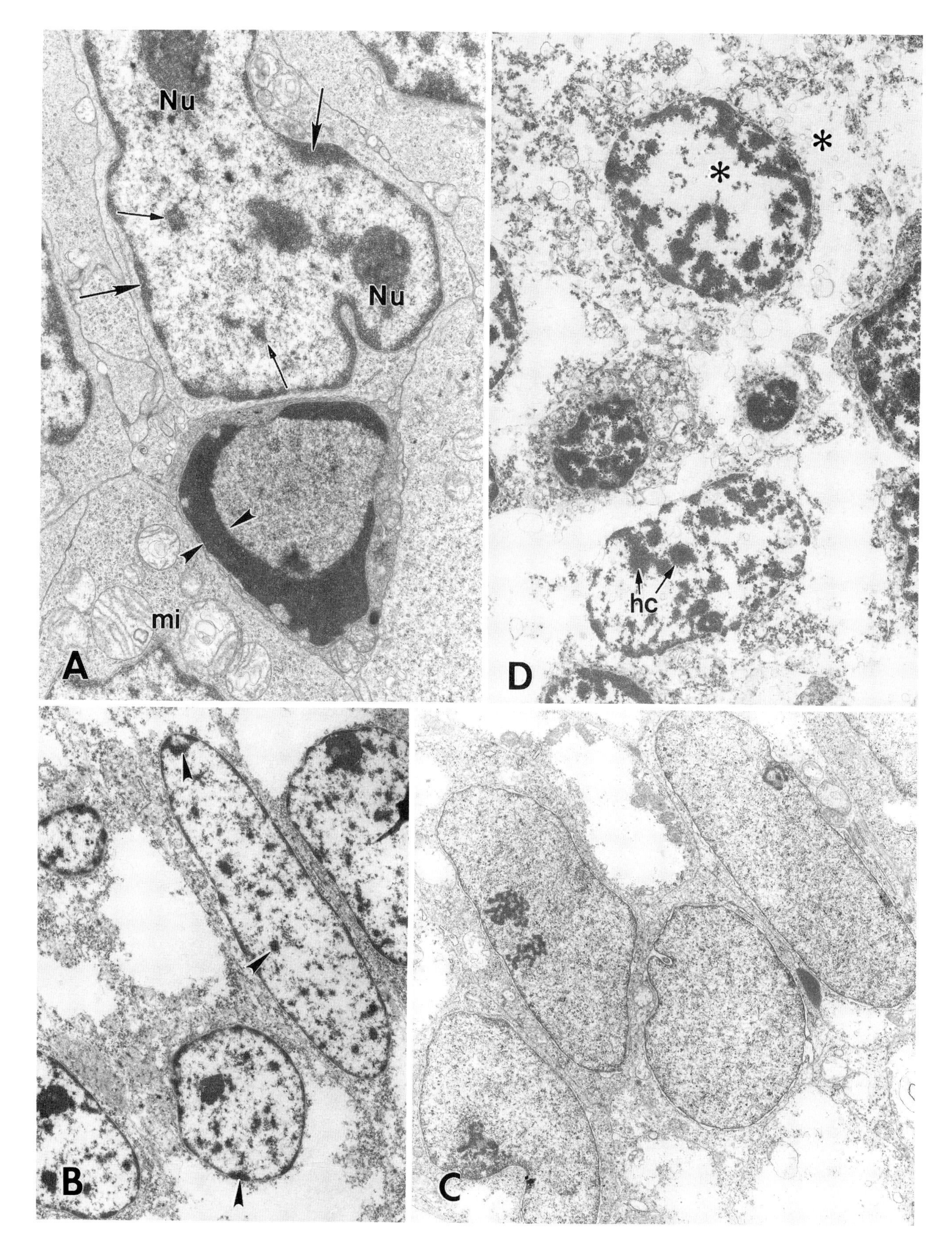
Nu
Nu
mi
A
*
*
hc
D
B
C

NUCLEAR ENVELOPE

The **nuclear envelope** consists of two membranes separated by a narrow clear space (Plate 2A; see also Plates 34A, 38A, and 63D): less commonly, it is referred to as the **perinuclear cisterna** because in effect it is a specialized example of rough endoplasmic reticulum (Chapter 3); ribosomes are attached to the outer surface and unambiguous rough endoplasmic reticulum is sometimes in continuity with it (Plate 2B).

The inner and outer membranes of the nuclear envelope meet at sites referred to as **pores**, which permit bidirectional transport of molecules between the nuclear interior and the cytoplasm (Plate 2A; see also Plates 34A, 38A, and 63D). The pore in cross section often has a distinctive bar across it, and heterochromatin is excluded from the immediate vicinity of the pore.

When the nuclear envelope surface lies at a small angle to the plane of the section, *en face* views are produced. These reveal pores to be circular (Plate 2C and 2D), and, depending on sectioning geometry, pores may be dense or haloed and *apparently* located in the cytoplasm (Plate 2C and 2D). Nuclear pores have sometimes been interpreted as virus-like particles. However, their exclusion from areas of nuclear envelope sectioned at 90° suggests that they are a structural component of the envelope.

NUCLEAR FIBROUS LAMINA

In vertical sections of the envelope, one often sees finely textured material of moderate density between the inner membrane of the envelope and the heterochromatin; this is the **nuclear fibrous lamina** (Plate 2A). It is compositionally similar to intermediate filament proteins (Chapter 24) and provides at least a means of physical contact between the nuclear envelope and the DNA itself. It is about 40–60 nm thick and is absent from the pore vicinity.

All cells–normal, reactive, and neoplastic—are believed to possess a nuclear fibrous lamina, although it varies in thickness. The nuclear fibrous lamina is:

- Best developed in mesenchymal cells
- Poorly developed in lymphoid and myeloid cells
- Often well-developed in Langerhans cells and melanocytes
- Variably developed in epithelial cells

NUCLEAR IRREGULARITY: NUCLEAR POCKETS AND PSEUDOINCLUSIONS

Normal cells typically have fairly smoothly outlined nuclear profiles, whereas—despite many exceptions—nuclear irregularity is frequently encountered in neoplastic cells and can assume varied appearances. Minor degrees of irregularity consist of shallow deformations of the nucleus by cytoplasm; these areas, when sectioned appropriately, can suggest a cytoplasmic profile within the nucleus. These only appear as intranuclear as artifacts of sectioning geometry, and they are therefore referred to as **nuclear pseudoinclusions** (Plate 3A).

Some structures resembling pseudoinclusions have a cytoplasmic composition different from the perinuclear cytoplasm (Plate 3B), and this hints at the possibility that the inclusion in question really is isolated from the surrounding cytoplasm. Only serial sectioning will

---

Plate 2. **A**: Nucleocytoplasmic interface defined by nuclear envelope, nuclear fibrous lamina (*large arrow*) and dense heterochromatin granules; some are in a row along the inner surface of the nuclear fibrous lamina (*arrowheads*); others (*) form an aggregate amounting to a piece of peripheral heterochromatin. Otherwise the nucleus is euchromatinic. A nuclear pore is present (*small arrow*). The cytoplasm contains recognizable rough endoplasmic reticulum cisternae and intermediate filaments. Mesenchymal chondrosarcoma, thigh. ×46,800. **B**: Nuclear envelope in continuity with unambiguous rough endoplasmic reticulum (*). The nucleus is euchromatinic and the rough endoplasmic reticulum is expanded by a moderately dense secretory product. Fibrosarcoma, subcutis, neck. ×20,500. **C**: Nucleus in which part is sectioned to reveal nuclear envelope (*arrows*) and part is tangentially sectioned, revealing nuclear pores *en face* (*arrowheads*). Diffuse high-grade non-Hodgkin's lymphoma, axillary node. ×20,500. **D**: Detail of nuclear pores *en face*. Cytoplasm is on the right (*), and the pore marked with a small arrow is at a superficial level of the nuclear envelope; it has an annular structure and a central granule. Other pores have haloes (*arrowheads*) and are at the level of the peripheral heterochromatin. Soft tissue sarcoma NOS, neck. ×71,500. **E**: Nuclear pocket enclosing a profile of cytoplasm (cy). Asterisk (*) indicates perinuclear cytoplasm. Note row of compressed heterochromatin granules (*arrow*). Anaplastic large-cell lymphoma, pelvis. ×69.000. **F**: Multiple projections of nuclear envelope producing an appearance of concentric nuclear pockets (*arrows*). Follicle center lymphoma (signet-ring cell variant), femoral node. ×19,700. hc, heterochromatin; ne, nuclear envelope; rER, rough endoplasmic reticulum.

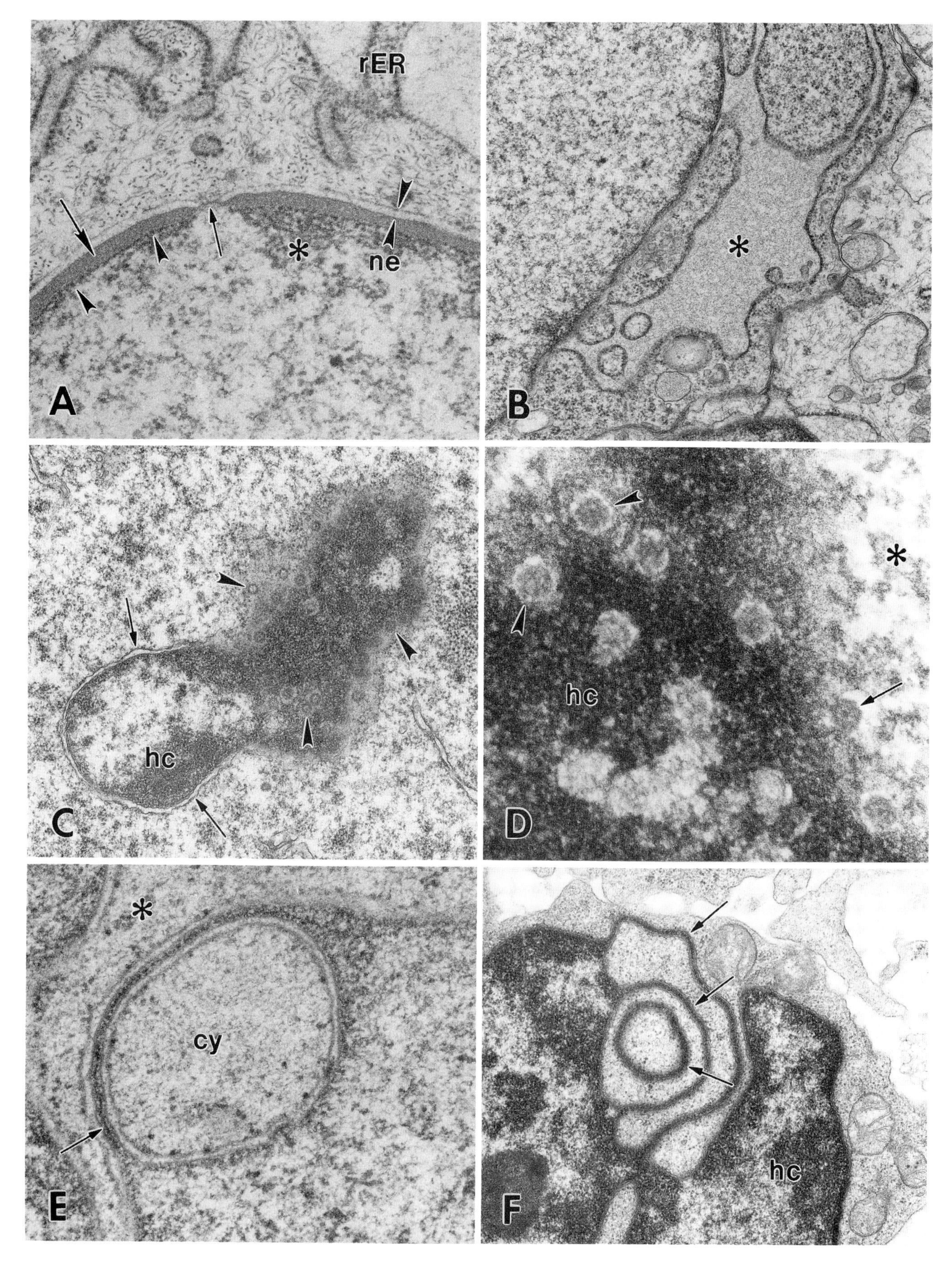
rER
ne
A
B
hc
C
hc
D
cy
E
hc
F

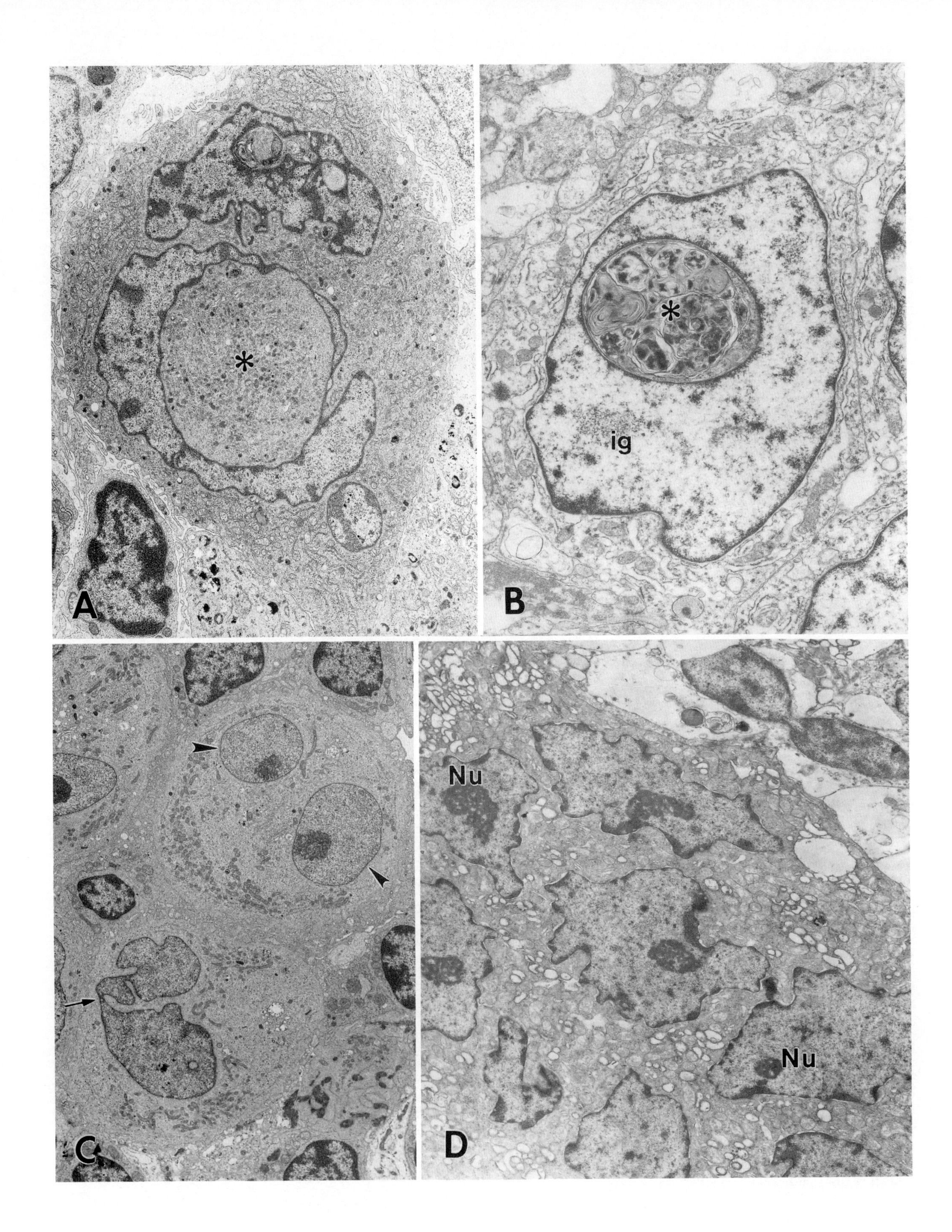
A
*
B
*
ig
C
D
Nu
Nu

unambiguously demonstrate this, however, since cytoplasmic composition can vary from one area of the cell to another (Dickersin et al., 1980).

A further common form of nuclear irregularity is the **nuclear pocket** (also referred to, less satisfactorily, as nuclear **blebs**). These seem to represent more localized pockets of cytoplasm protruding into the nucleus (Plate 2E). Mostly, the pockets contain cytoplasm of the same composition as the perinuclear cytoplasm; some, however, contain collections of clear vesicles (Ioachim, 1985). These pockets are also regarded as pseudoinclusions, but again only serial sectioning can demonstrate this. In addition, some pockets contain intranuclear rather than cytoplasmic material.

Nuclear pockets are present in normal centroblasts and can often be found in large numbers in non-Hodgkin's lymphomas (Plate 2E and 2F); in small numbers they are nonspecifically distributed among tumor cells.

NUCLEAR IRREGULARITY: MULTILOBATION AND MULTINUCLEARITY

One kind of nuclear irregularity takes the form of **multilobation**. Here, multiple nuclear profiles are present within the same cell, often connected to each other by fine bridges. Therefore, they represent lobes of a single nucleus (Plate 3C). The combination of nuclear bridges and the light microscope appearance of physically related nuclear profiles are important for interpreting multilobation (Elavathil et al., 1989). In **multinuclearity**, by light microscopy one sees nuclear profiles separated from one another; and by electron microscopy, bridges are absent (Plates 3D and 17C).

Excessive nuclear irregularity produces images of nuclear profiles abundantly dissected by cytoplasmic channels (Plate 4A). In most cases, these profiles belong to a single highly irregular nucleus and are connected by fine bridges consisting of a duplicated nuclear envelope with intervening heterochromatin.

Multilobated nuclei are found in:

- Neutrophils and some myeloid leukemias
- diffuse large B-cell lymphomas (multilobated subtype)

Multinuclearity characterizes:

- Multinucleated osteoclastic giant cells
- Osteoclast-like giant cell tumors

DISCRETE INTRANUCLEAR STRUCTURES

Apart from heterochromatin, the main components in the normal nuclear interior are the **nucleolus, perichromatin granules**, and **interchromatin granules**. Nucleoli are sites of synthesis of ribosomal RNA and of its assembly with protein into preribosomal particles; perichromatin granules and interchromatin granules contain protein and RNA and have less-well-defined functions.

NUCLEOLUS

Nucleoli are usually recognizable as discretely delineated structures. They vary in size as follows: They can be as large as several micrometers across, filling out a substantial proportion of the interchromatin; or they can be just a few hundred nanometers in diameter, appearing rather inconspicuous (see Plates 4, 11, 17, 20, 21, 26, 31, 33–36, 40, 42, 43, 45, 47, 48, 54, 56, and 60; see Plate 51B for a particularly small nucleolus). They stand out most readily in euchromatinic nuclei (Plates 4C and 60), but even in nuclei containing significant quantities of heterochromatin, they can usually be distinguished by a number of features.

They typically contain pale-staining rounded profiles of very homogeneous material (Plate 4B) called **fibrillar centers**. These are often enveloped in a complete or partial rim of highly compacted dense material referred to as the **dense fibrillar component**, best identified at

Plate 3. **A**: Pseudoinclusion (*) of large size (6 μm across). Note compositional similarity to perinuclear cytoplasm. Invasive malignant melanoma, scapular skin. × 5900. **B**: Large cytoplasmic profile (*) (4 μm across) where composition is different from that of perinuclear cytoplasm. It contains myelin figures and secondary lysosomes, whereas the surrounding cytoplasm is pale-staining and contains mitochondria and rough endoplasmic reticulum. The nucleus contains interchromatin granules. Schwannoma, soft tissues of iliac crest. x10,100. **C**: Nuclear lobation. One cell has two discrete nuclear profiles (*arrowheads*), while an almost identical cell shows a nucleus with a narrow bridge (*arrow*). Diffuse large B-cell lymphoma (multilobated variant), cervical node. × 3500. **D**: Multiple nuclear profiles not connected by nuclear bridges. Reactive multinucleated giant cell, plasmacytoma of maxillary antrum. × 5600. ig, interchromatin granules; Nu, nucleolus.

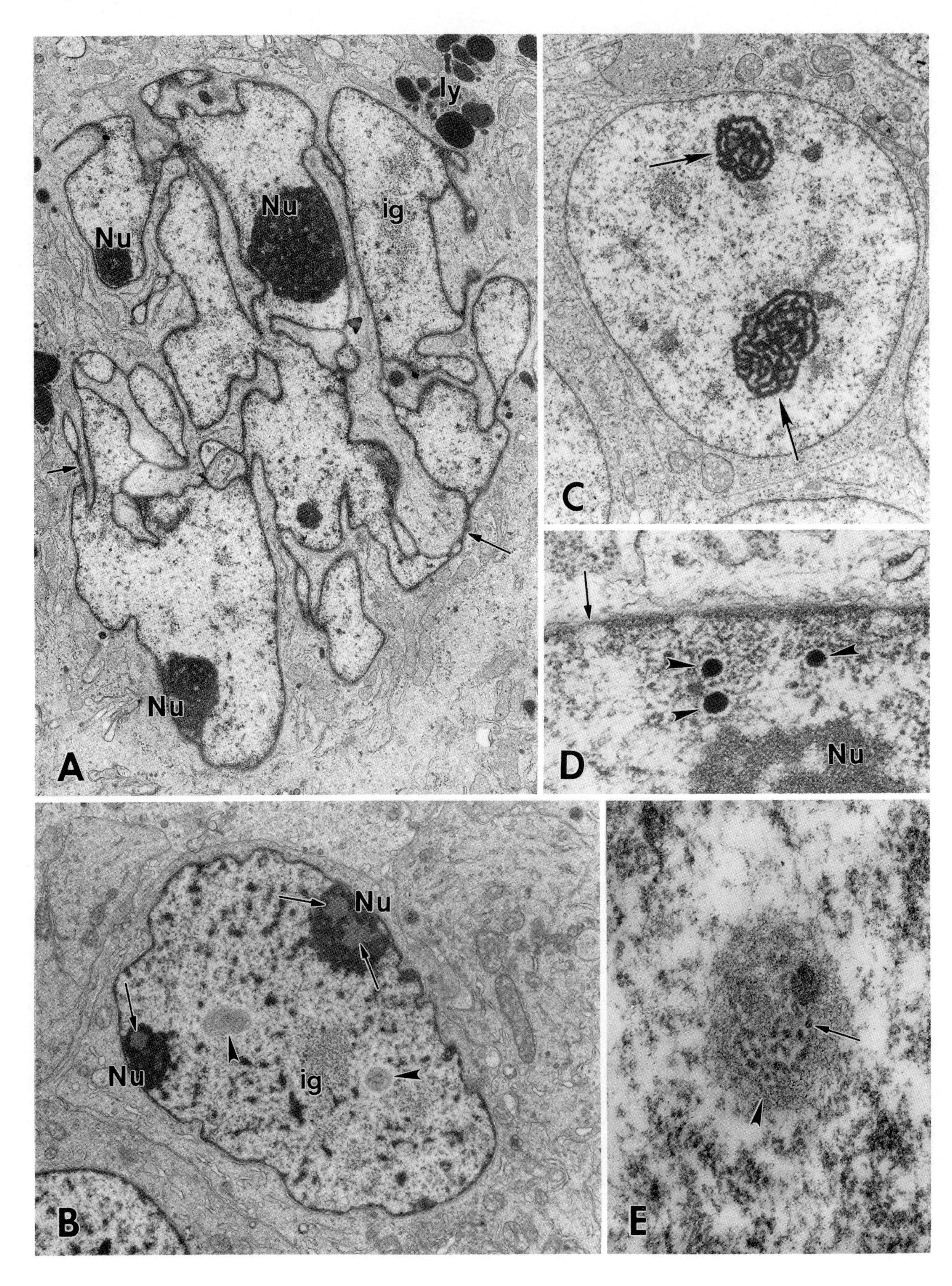

Nu
Nu
ig
ly
Nu
A
C
D
Nu
Nu
Nu
ig
B
E

high magnification. In addition, a **granular component** may be present, which contains pre-ribosomes.

Active nucleoli have a mix of these components. A significant predominance of one indicates a defective or an inactive nucleolus; for example, nucleoli with a uniform texture or consisting exclusively of thread-like profiles are probably inactive or defective. Marginal nucleoli—attached to the nuclear envelope—are indicative of cells actively involved in protein synthesis (Plates 4, 33, 47, and 60).

Some nucleoli are distinguished by rope-like profiles of dense material, referred to as **nucleolonemata** (singular: **nucleolonema**): these nucleoli are frequently found in seminomas (Plate 4C).

In some cells and under certain staining conditions, nucleoli have a distinctly lesser or greater density compared with the surrounding **nucleolus-associated heterochromatin** (Ward et al., 1981; Tykocinski et al., 1984).

PERICHROMATIN GRANULES: INTERCHROMATIN GRANULES

Perichromatin granules are sharply delineated, rounded, dense structures found either at the heterochromatin–interchromatin interface or within the heterochromatin. They have a fine halo and usually measure about 20 nm in diameter (Plates 29A and 41A). **Interchromatin granules** are smaller (10–20 nm in diameter) and are found in aggregates in the interchromatin (Plate 4A and 4B).

Perichromatin and interchromatin granules are nonspecifically distributed amongst normal and tumoral cells, but particularly large examples (~300 nm) of perichromatin granules (Plate 4D) are characteristic of nasopharyngeal angiofibromas.

OTHER INCLUSIONS

In tumor cell nuclei, in addition to these structures, one may see a variety of other organelles. Many of these are similar to, and may derive from, structures normally found in the cytoplasm or extracellular matrix; some may result from abnormal synthesis in the interchromatin. Examples include: smooth endoplasmic reticulum (e.g., the **nucleolar canalicular structures** typical of endometrium and type II pneumocytes); structures resembling annulate lamellae; melanosomes; myelinosomes; crystals; lipid; glycogen; vacuoles; filaments; viruses; whole cells; nonspindle microtubule aggregates; immunoglobulin (Dutcher bodies); and foci of calcification.

Some inclusions are exclusively intranuclear with no cytoplasmic counterpart. They include several type of body referred to as **true intranuclear inclusions**, which are recognized as existing in four morphologically distinct variants (Duquene and Dourov, 1991). Type III (Plates 4E and 5D) is commonly encountered and seems to have a nonspecific distribution.

Plate 4. **A**: Single highly irregular nucleus in which nuclear profiles are joined by nuclear envelope bridges (*arrows*). The nucleus is mostly euchromatinic and contains several nucleoli and a focus of interchromatin granules. Interdigitating cell sarcoma, recurrent in jejunum. ×8500. From a block supplied by Dr. S. Nakamura (Nagoya, Japan). **B**: Nucleus with a speckled heterochromatin pattern, two nucleoli with fibrillar centers (*arrows*), two true intranuclear bodies (*arrowheads*), and a group of interchromatin granules. Gastrointestinal autonomic nerve tumor, abdomen. ×8400. **C**: Essentially euchromatinic nucleus with two reticulate nucleolonematous nucleoli (*arrows*). Seminoma, metastatic to humerus. ×6100. **D**: Slightly larger than usual perichromatin granules (*arrowheads*). Note also the position of nuclear pores and clearing of adjacent heterochromatin (*arrow*). Epithelioid leiomyosarcoma, peritoneum. ×39,900. **E**: True intranuclear body (type III) showing coarse granules (*arrow*) and finely textured filamentous material (*arrowhead*). Reactive myofibroblast from nodular sclerosing Hodgkin's disease. ×76,000. ig, interchromatin granules; ly, lysosomes; Nu, nucleolus.

3

Endoplasmic Reticulum

Endoplasmic reticulum was the name originally given to the membranous system found in the central area of cytoplasm ("endoplasm") in cultured cells. The term now applies to membranes with distinctive ultrastructural appearances found throughout the cell cytoplasm. The term *ergastoplasm* is obsolete.

There are two morphologically distinct types: (1) **Rough** or **granular endoplasmic reticulum** has ribosomes attached to the cytosol-facing surface; (2) **smooth endoplasmic reticulum** lacks ribosomes and is therefore smooth-membraned.

ROUGH ENDOPLASMIC RETICULUM

Rough endoplasmic reticulum (rER) is the site of synthesis of proteins for export through the Golgi apparatus and secretory granules. Cells and tumors vary considerably in their rER content. Synthetically inactive cells may have only one or two slender cisternae, whereas numerous rER cisternae suggest synthesis of proteinaceous secretions such as neuroendocrine, lysosomal and serous granules, immunoglobulin and matrix proteins.

Abundant rER is diagnostically important in identifying the following tumors (and their normal cell counterparts): plasmacytomas, myelomas, lymphoplasmacytoid lymphomas; adenomas and adenocarcinomas of salivary gland and exocrine pancreas (e.g., acinic cell tumors); neuroendocrine and steroidogenic tumors (e.g., pituitary adenomas, adrenocortical adenomas and carcinomas); fibroblastic lesions including fibromas of various kinds (e.g., fibroma of tendon sheath), fibrosarcomas, nodular fasciitis, myofibromatoses, reactive myofibroblasts, myofibrosarcomas, malignant fibrous histiocytomas; cartilaginous and osseous tumors including osteosarcomas, chondromas, chondrosarcomas (including the extraskeletal myxoid type).

For cells containing conspicuous rER, see Plates 5, 17A, 19, and 29A.

Plate 5. **A:** Abundant rER (*large arrows*) in the form of cisternae punctuated by cytoplasm (*small arrows*). The large dense bodies are probably lysosomes. Fibrolamellar hepatocellular carcinoma, metastatic to celiac node. ×9100. **B:** A small stack of rER cisternae cut at right angles and appearing as tubules; two (*arrowheads*) are continuous with areas showing ribosomes *en face* (*). Aggressive angiomyxoma, broad ligament. ×60,000. **C:** Cleft-like spaces (*arrows*) in a cell containing abundant rER. Osteogenic sarcoma, carotid bifurcation. ×7920. **D:** Detail from **C** showing cleft (*) in continuity with rER (*arrow*). ×60,800. **E:** Several vesiculated rER cisternae (*small arrows*) in a cell also showing poorly preserved (i.e., distended) nuclear envelope (*large arrow*). Malignant fibrous histiocytoma, parotid. ×19,400. ly, lysosomes; N, nucleus.

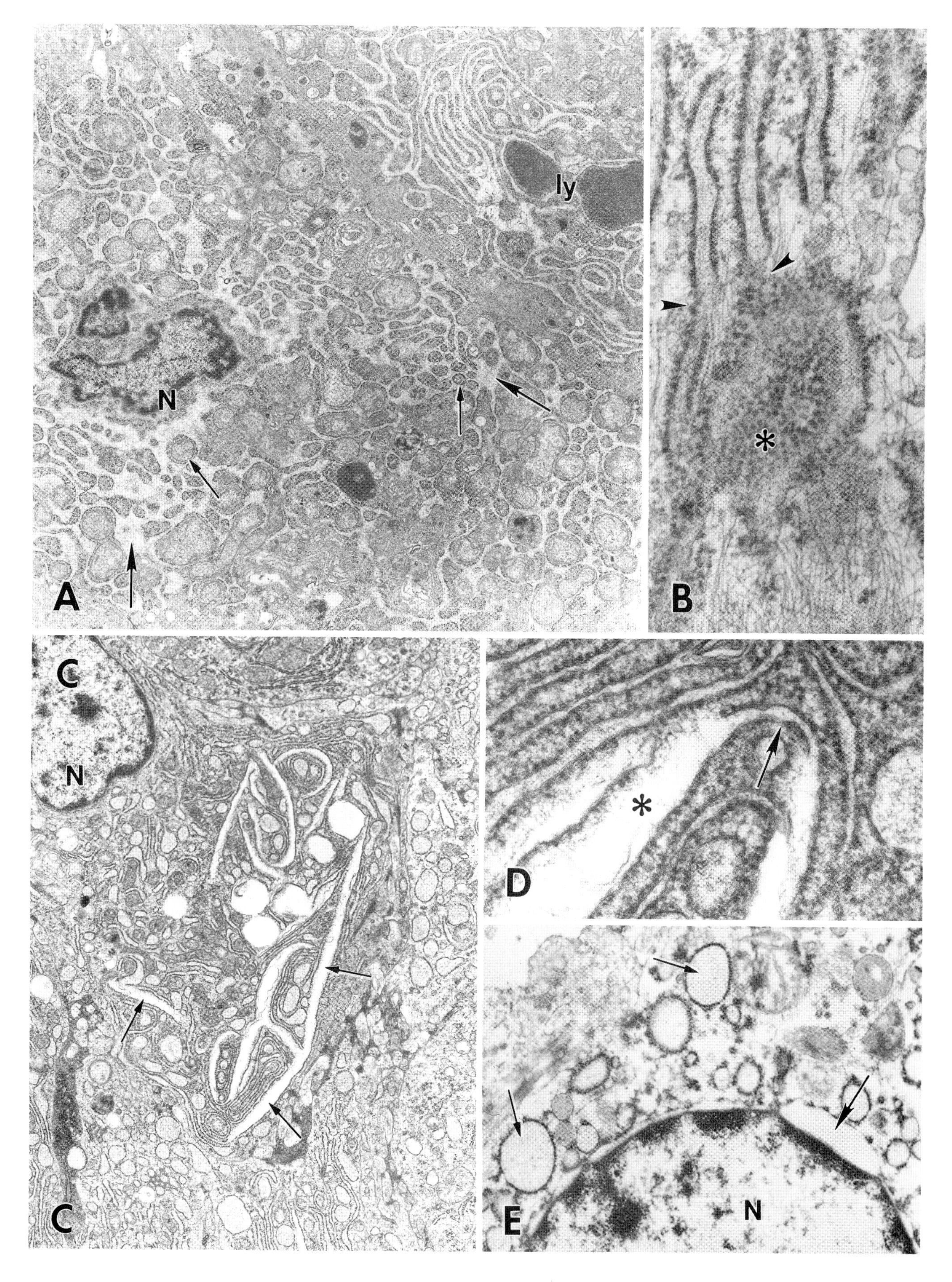
ly
N
A
B
C
N
C
D
E
N

ULTRASTRUCTURE OF rER

rER typically consists of plate-like cisternae of ribosome-studded membrane; when this lies in the plane of the section and is punctuated by cytoplasm, it resembles a reticulum (Plate 5A). When such cisternae lie at 90° to the plane of the section, they appear in profile as tubules. Sometimes the rER surface undulates in and out of the plane of the section, producing *en face* views: here, a gray expanse of membrane with attached ribosomes is observed (Plates 5B and 58A).

In terms of arrangement, cisternae may lie in parallel, sometimes forming short stacks. They may be angular, and large numbers of such cisternae may form whorled structures. A specialized cisternae of unilaterally ribosome-studded endoplasmic reticulum is associated with the Golgi apparatus (Plate 9). rER also forms complexes with mitochondria (Plate 52C).

Distribution of Distinctive rER Arrangements

- Short stacks of rER are often found in adrenocortical, hepatocellular and neuroendocrine tumors.
- Prominent rER–mitochondrion complexes have been seen in medullary thyroid carcinomas, chief cell parathyroid adenomas, pleomorphic adenomas, hepatocellular carcinomas, chordomas and malignant fibrous histiocytomas.

Artifactual appearances include electron-lucent clefts and sphericalization of rER cisternae. Clefts can appear in reasonably well-preserved cells (Plate 5C and 5D) and occur in mesemchymal and epithelial cells. Sphericalization (vesiculation) of rER cisternae (Plate 5E) is typically an indication of poor preservation. Small sphericalized rER cisternae have been misinterpreted as secretory vacuoles—especially under low magnification, where the ribosomes may not be very conspicuous. Large ones can be confused with the vacuoles forming primitive lumina in vascular cells and thereby promote an erroneous interpretation of angiosarcoma. rER can partially degranulate and mimic smooth endoplasmic reticulum, and it can be seen in tissue processed from paraffin wax.

THE INTERIOR OF rER CISTERNAE

rER cisternae possessing a tubule-like profile in cross section often have a fairly clear content and contain only a small amount of very finely textured electron-dense material (Plate 5B). This appearance reflects normal function in indicating that newly synthesized peptide is being continuously transferred to the Golgi apparatus; this avoids an intracisternal buildup of protein. Very often in tumors, however, one finds evidence of a failure to transport protein, in which rER cisternae become distended (Plates 2, 6 and 8). Sometimes these distended cisternae are large enough to be visible by light microscopy (Callea et al., 1986). Often, accumulated protein is very homogeneously textured and fairly light-staining (Plate 2B). It may also contain foci of denser material (Plate 6A).

Morphologically distinct intracisternal inclusions with characteristic lesional distributions include the following: Russell bodies (immunoglobulin) in plasma cells and plasma cell proliferations (Plate 6B); paracrystalline immunoglobulin in certain classes of lymphoma and leukemia; paracrystalline inclusions of unknown nature and described as "serpigenous" in malignant schwannomas, angiosarcomas, fibroblastic proliferations, epithelioid sarcomas and prostatic adenocarcinomas; fine filaments (putative collagen); microtubular inclusions (Plate 6C and 6D) in malignant melanomas, myxoid chondrosarcomas, chordomas, osteosarcomas, basal cell adenomas (stromal cells), small bowel adenocarcinomas, rhabdoid tumors and ependymomas; tubuloreticular structures in endothelium and/or blood lymphocytes in a variety of conditions: acquired immunodeficiency syndrome (AIDS), lupus erythematosus, myelodysplastic syndrome, AIDS-related and non-AIDS Kaposi's sarcomas, gliomas, germinomas.

The microtubular inclusions (Plate 6C and 6D) have a diameter of ~15–30 nm; this puts them roughly within the size spectrum for true tubulin-containing microtubules (Chapter 25). There is, however, no consistent evidence that these structures really consist of tubulin. Suzuki et al. (1988) obtained positive staining of extraskeletal myxoid chondrosarcoma with antitubulin

Plate 6. **A:** An expanded rER cisterna containing dense foci (*arrow*) embedded in lighter-staining proteinaceous material (*). Note secondary lysosomes. Malignant granular cell tumor, scalp. ×21,100. **B:** Spherical rER cisterna containing aggregate of highly electron-dense inclusions (*small arrow*) producing medusa-form Russell body. Note the solitary small inclusion (*large arrow*) in its own cisterna. Polypoid lymphoid mass, rectum. ×14,600. **C** and **D:** Intracisternal microtubular inclusions (*arrows*). Note the internal density in two of the inclusions in **D** (*arrowheads*). Extraskeletal myxoid chondrosarcoma of thigh (**C**) and buttock (**D**). **C**, ×67,300; **D**, ×146,000. **E:** TRSs (*small arrows*) in a cultured lymphoblastoid cell also showing cylindrical confronting cisternae (*large arrows*). ×48,000. ly, lysosome; N, nucleus; Nu, nucleolus. [Micrograph courtesy of Samuel Hammer, M.D. (Seattle, Washington). Reproduced from Bockus et al. (1988), with permission from WB Saunders Company.]

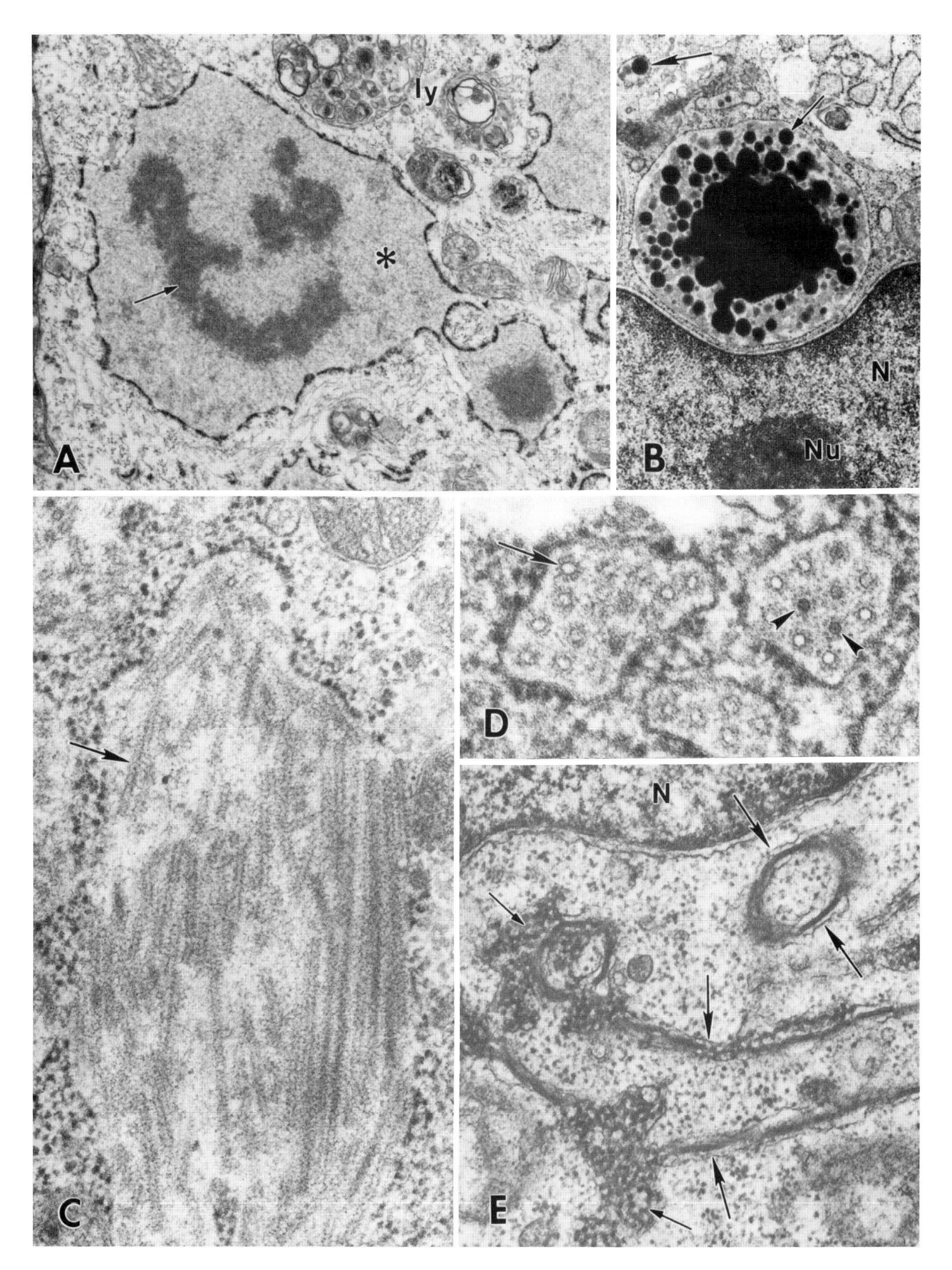
ly
*
A
B
N
Nu
C
D
N
E

antibodies by light microscopy, but Hearn and Kontozoglou (1989) failed to obtain immuno-ultrastructural localization in a malignant melanoma with abundant similar intracisternal inclusions. Therefore, it is preferable to refer to these structures as **microtubular inclusions** rather than **microtubules.** The larger-diameter tubules (70–90 nm across) observed in chronic lymphocytic leukemia are probably even less likely to be of tubulin, and their composition is also undefined (Kostianovsky and Ghadially, 1987).

Tubuloreticular structures (TRSs; formerly also called **tubuloreticular inclusions,** TRIs) form a reticulum of 20 to 30-nm-diameter tubules within rER cisternae (Plate 6E). In some early studies their similarity to myxoviruses was pointed out (Gyorkey et al., 1971). The TRS cisternal membrane has been shown on some occasions to be continuous with the membranes of **cylindrical confronting cisternae** and **annulate lamellae** (see below). TRSs are important because they are present in nearly all blood lymphocytes from AIDS patients, although they are not specific for this condition. The paper by Grimley and Schaff (1976), although old, gives a more detailed list.

SMOOTH ENDOPLASMIC RETICULUM

Smooth endoplasmic reticulum (sER), especially when found with lipid and mitochondria, is implicated in lipid, glycogen and steroid metabolism. **Sarcoplasmic reticulum,** the specialized sER of skeletal muscle, has a role in controlling calcium ion concentrations inside and outside the cell, which, in turn, controls contraction.

sER is often abundant and therefore important in the diagnosis of steroidogenic and liver cell tumors—for example, adrenocortical adenomas and carcinomas, Leydig cell tumors, ovarian lipid cell (steroid cell) tumors, lutenizing thecomas, granulosa cell tumors, hepatomas, hepatocellular carcinomas. It is also present in some neuronal tumors (e.g., gastrointestinal autonomic nerve tumors—Plate 7C), lymphomas and leukemias, renal cell carcinomas, neuroendocrine tumors (e.g., pituitary adenomas) and schwannomas.

ULTRASTRUCTURE OF sER

In steroidogenic cells, well-preserved sER consists of a reticulum of tubules with rather delicate-looking membrane. Often, however, the reticulum is disrupted by fixation and one sees individual tubules or vesicles. These can be either very closely packed, filling the spaces between organelles such as lipid droplets and mitochondria (Plate 7A), or somewhat more separated from one another and rounded or angular (Plate 7B); here, nevertheless, one can sometimes see remnants of the reticulum (Gaffney et al., 1983). Unlike rER, sER does not have internal dense material. sER may also be organized concentrically: such an arrangement based on a central lipid droplet is recognized as a **spironolactone body** and is found in adrenocortical cells (see Favre et al., 1980).

SARCOPLASMIC RETICULUM

Sarcoplasmic reticulum (Plates 53A and 54C) is physically related to the myofibrils of striated muscle; in conjunction with the T-tubule system, it forms the so-called **triads.** For further details of triads, see Cornog and Gonatus (1967) and Schlosnagle et al. (1983).

Smooth-membraned systems encountered in reactive and neoplastic cells but not generally referred to as smooth endoplasmic reticulum include: smooth-membraned aggregates in acute leukemia (Seman, 1981); tubules associated with the plasmalemma in alveolar soft-part sarcoma [may be related to endocytosis or mimic the T-tubule system of striated muscle (Ohno et al., 1994)]; curvilinear bodies (worm-like bodies; comma-shaped bodies) found in cells with a hint of histiocytic or lymphoid differentiation (Plate 7D); intranuclear smooth-membraned systems in type II pneumocytes and endometrium.

Plate 7. **A:** Compact tubules and vesicles (*arrows*) of sER completely filling the spaces between lipid droplets. Steroid cell tumor, ovary. ×60,800. **B:** Rounded and angular sER cisternae (*arrows*). Some rER cisternae are present, possibly giving rise to sER (*arrowhead*). Leydig cell hyperplasia associated with adenocarcinoma metastatic to ovary. From a block courtesy of Irving Dardick, M.D. (Toronto). ×53,500. **C:** A reticulum of smooth membranes (*arrows*) embedded in intermediate filaments. Gastrointestinal autonomic nerve tumor, peritoneum. ×53,500. **D:** Curvilinear body in the form of an undulating plate of smooth membrane. The gray areas (*arrows*) represent membrane *en face*. Although the organelle is very largely smooth, some indications of contact with rER have been demonstrated (Nakamura et al., 1989). Similar bodies can be found within lysosomes. Hodgkin's disease, axillary node. ×53,500. if, intermediate filaments; li, lipid; pm, plasmalemma; rER, rough endoplasmic reticulum.

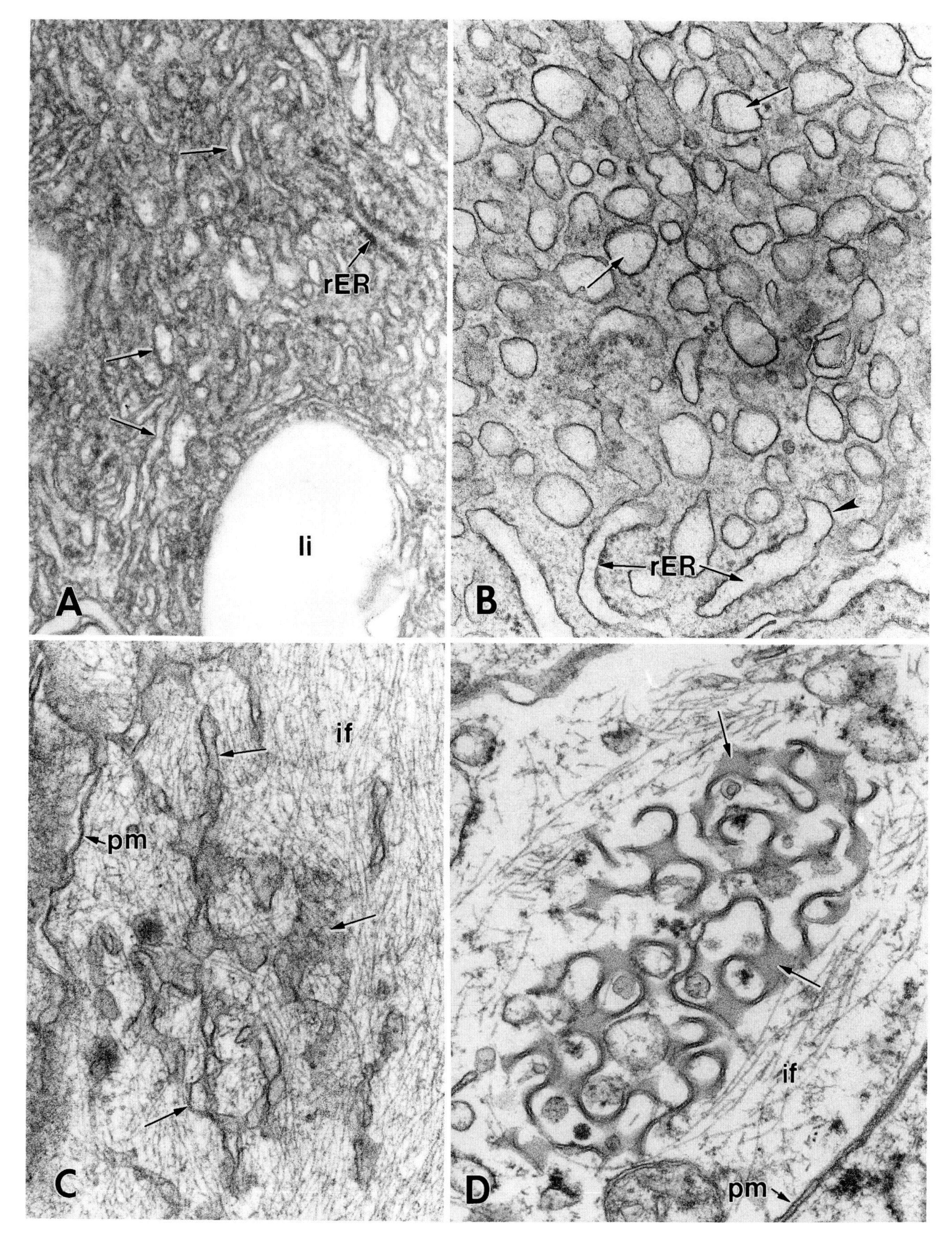
rER
li
A
rER
B
if
pm
C
if
pm
D

ORGANELLES COMBINING SMOOTH AND ROUGH MEMBRANE

A variety of membrane systems are encountered in tumors combining smooth and rough membrane. Under normal circumstances, sER appears to be generated in some cells from rER, and one can occasionally see unambiguous rER focally transformed into sER. One can also encounter loops or whorls of membrane which are recognizable as rER only by virtue of an occasional residual ribosome (Sohval et al., 1982). A further example is the cisterna closely apposed to the vacuole in vacuolar signet-ring cell lymphoma; this is mostly smooth but retains sparsely distributed ribosomes (Eyden et al., 1990).

Several other distinctive categories of smooth-and-rough endoplasmic reticulum are known. In many instances, these lack diagnostic value but it is nevertheless important to identify them for what they are and not to misinterpret them as diagnostic structures.

CONFRONTING CISTERNAE

Confronting cisternae consist of pairs or larger numbers of rER cisternae arranged in parallel such that the apposed surfaces are devoid of ribosomes (Plate 8A). Some appear to be remnants of the nuclear envelope and are seen during and after mitosis (Plate 8A and 8B); others seem to be present in tumor cells without evidence of cell division (Nakanishi et al., 1986).

Cylindrical confronting cisternae are a variant of confronting cisternae with a distinctive tubular or concentric profile (Plate 6E) and are believed to have a test-tube shape. Like tubuloreticular structures, they are found in blood lymphocytes of AIDS patients (Ghadially et al., 1987; Kostianovsky et al., 1987).

ALVEOLATE RETICULUM

Organelles are also encountered which are similar to confronting cisternae in having "outer" ribosomally coated membrane and "inner" smooth membrane associated with dense material, but which are structurally more complex (Plate 8C). On some occasions, such structures have been given an inappropriate terminology; for example, they have been described as **tubuloreticular structures,** but this term is already established for a distinctive intracisternal organelle (Plate 6E). The purely descriptive term **alveolate reticulum** is probably best (Uzman et al., 1971).

STELLATE rER

Stellate rER is a variant of alveolate reticulum in which the peripheral rER cisternae have a radiating pattern (Plate 8D). Stellate rER has been associated particularly with promyelocytic leukemia (Plate 8D).

Other variants of "smooth-plus-rough" endoplasmic reticulum exist with striking geometrical arrangements. Some of them have complex and sometimes confusing names and have no known diagnostic importance, but they are recorded here for completeness: undulating tubules associated with endoplasmic reticulum (Chandra, 1968), tubuloampullar structures (Graf and Heitz, 1980), cytoplasmic tubular inclusions (Matsuda and Nagashima, 1984), intracisternal microtubular reticular structures and crystalline microtubular inclusions (Szakacs et al., 1991), endoplasmic reti-

Plate 8. **A:** Part of a mitotic cell containing heterochromatin surrounded by a nuclear envelope which in part is duplicated in the form of simple confronting cisternae (*arrows*). Rhabdomyosarcoma, cheek. ×11,600. **B:** Detail of confronting cisternae in a mitotic cell. One pair (*arrow*) is continuous with an expanded cisterna of rER (*); another (*arrowheads*) is attached to heterochromatin. The cisternae seem to be cemented together by a finely textured dense material. Malignant schwannoma, breast. ×83,300. **C:** Alveolate reticulum with outer rER (*large arrows*). Inner cisternae (imaged as vesicles or tubules) of smooth membrane (*small arrows*) are embedded in dense amorphous material. Stromal sarcoma, breast. ×46,300. **D:** Ribosome–lamella complexes (*large arrows*). Note the lamellae (*arrowheads*). Hairy cell leukemia. ×28,500. [Micrograph courtesy of Professor M. Djaldetti (Petah-Tikva, Israel). Reproduced from Djaldetti (1976), with permission from Philips Electron Optics (Eindhoven, The Netherlands). An enlarged version of the micrograph was used as Figure 2 in F. N. Ghadially, *Ultrastructural Pathology and the Cell and Matrix*, 3rd edition, Butterworth–Heinemann, Ltd. (Oxford, England).] **E:** Stellate rER. Note focus of smooth-membraned elements surrounded by dense material (*large arrow*) and "radiating" rER cisternae (*small arrows*). Lymphoplasmacytoid lymphoma, inguinal node. ×57,000. **F:** Concentric annulate lamellae showing continuity with unambiguous rER. Note pores (*arrows*). Pleomorphic/myxoid malignant fibrous histiocytoma, subcutis, thigh. ×47,800. hc, heterochromatin; mi, mitochondrion; N, nucleus; rER, rough endoplasmic reticulum.

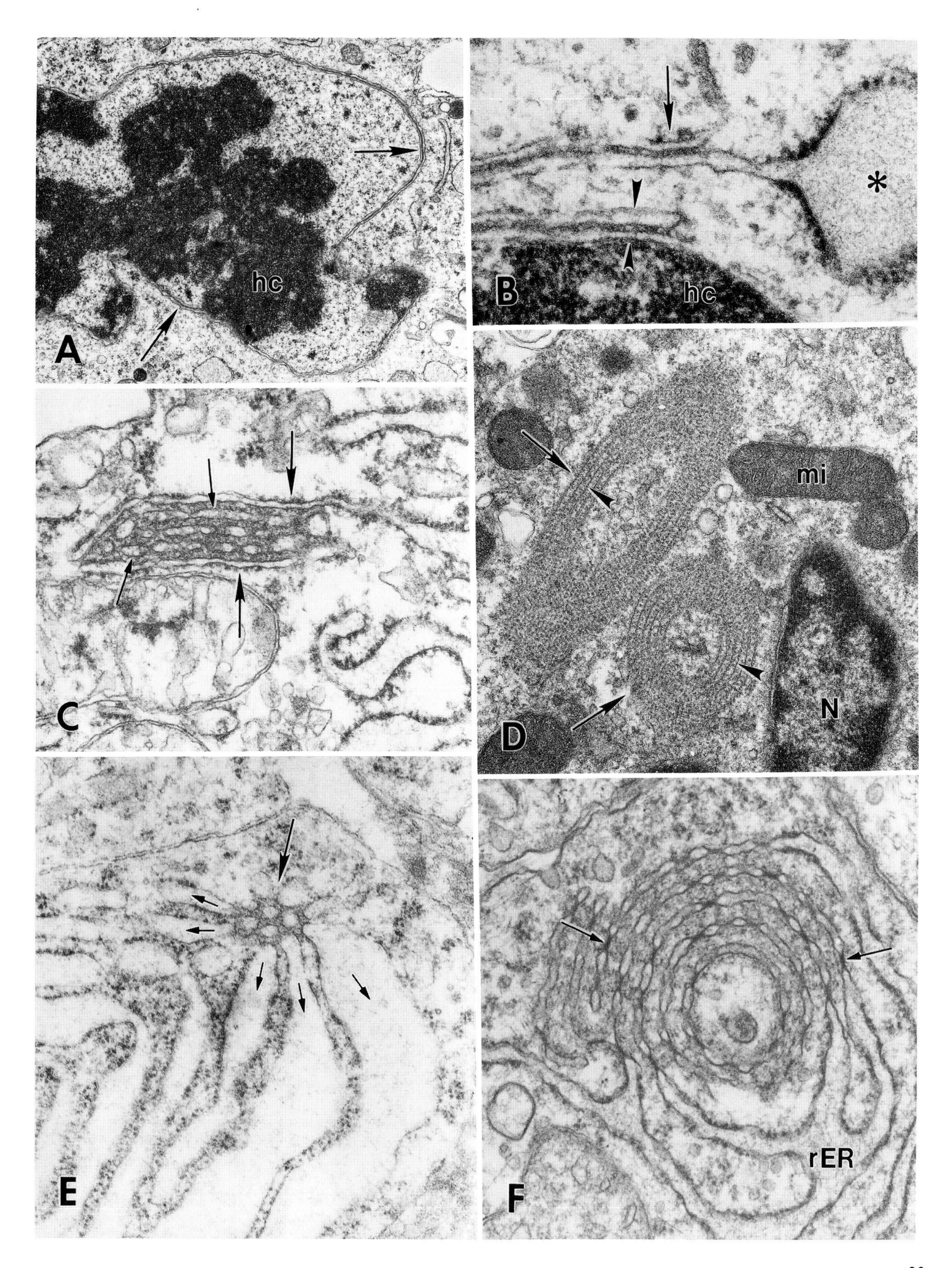
A
hc
B
hc
*
C
D
mi
N
E
F
rER

culum with honeycomb appearance (Gompel, 1971; Carstens et al., 1984), lamellar bodies (Tokue et al., 1985), concentric membranous bodies (Ellis and Coaker, 1989) and vesicular rosettes (Eyden et al., 1993).

RIBOSOME–LAMELLA COMPLEXES

Ribosome–lamella complexes (RLCs) are cylindrical structures consisting of concentrically arranged lamellae to which ribosomes are attached* (Plate 8E). Older terms encountered in the literature but not presently used include: *polysome–lamellae complex, granulofilamentous body and particle–lamella complex.*

RLCs were originally associated with hairy cell leukemia, but they have also been identified in a wide range of neoplastic conditions, of which the following are some examples: hairy cell, B-cell chronic lymphocytic and monoblastic leukemias, Sézary's syndrome, myeloma, lymphoma; reactive plasma cells, mast cells, lymphocytes and sinus histiocytes; parathyroid adenoma, pituitary adenoma, paraganglioma, insulinoma; Sertoli cell tumor; adrenocortical adenoma, pulmonary adenocarcinoma; ossifying fibromyxoid tumor of soft parts, hibernoma, alveolar soft-part sarcoma; ganglioglioma, astrocytoma, meningioma, hemangioblastoma.

* Strictly speaking, RLCs do not consist of rough and smooth membrane, but they are included here for convenience because their lamellae with attached ribosomes superficially resemble rER.

ANNULATE LAMELLAE

Annulate lamellae consist of cisternae bearing pores similar to those of the nuclear envelope (Plate 8F). The pores often appear in rows or other geometric arrays. Typical bar-like structures can be seen traversing the pores, while *en face* views from tangentially sectioned areas reveal circular outlines. Annulate lamellae may consist of straight parallel cisternae or have a concentric pattern (Plate 8F). Intranuclear annulate lamellae have also been observed.

Some smooth-membraned structures (e.g., honeycomb endoplasmic reticulum) have been misinterpreted as annulate lamellae (Cooney et al., 1982).

Annulate lamellae have been observed in germ cells, Sertoli cells and embryonic cells, pituitary and parathyroid adenomas and a variety of malignancies (carcinomas, lymphomas, melanomas, sarcomas).

4

Golgi Apparatus

The **Golgi body** (or **Golgi apparatus**) (Plate 9A) has the role of receiving newly synthesized protein from the rough endoplasmic reticulum (rER), altering it biochemically (for example, by sulfation and glycosylation) and packaging it as membrane-bound secretions.

The Golgi apparatus is relatively insignificant diagnostically compared with the secretory products it helps to form, but it is often well developed in cells and tumors actively synthesizing and secreting proteinaceous materials—for example, plasma cells, plasmacytomas, myelomas; pancreatic and salivary epithelium, acinar cell carcinomas (Plate 19): neuroendocrine cells and tumors; fibroblasts and myofibroblasts and their proliferations; chondroblasts, chondrosarcomas, osteoblasts, osteosarcomas.

A note on terminology: **Golgi body** or **Golgi apparatus** indicates a single discrete stack of cisternae (often called **saccules**). **Golgi complex** can be used for the entire Golgi complement, whether this is a single Golgi body or a large number. **Dictyosome** also applies to a single stack of Golgi saccules but has a more botanical usage.

ULTRASTRUCTURE OF THE GOLGI APPARATUS

The Golgi apparatus is an assembly of several aligned and flattened smooth-membraned saccules, frequently found near the nucleus (Plate 9A and 9C). The Golgi apparatus has several components or associated structures. An adjacent rER cisterna, lacking ribosomes over the surface facing the Golgi saccules (Plate 9A and 9C), produces **transitional** or **intermediate vesicles** (Plate 9A) which transport protein to the Golgi saccules. A stack of flattened saccules develops as a result of the fusion of transitional vesicles with the convex surface of the stack. Other vesicles, either clathrin-coated or smooth, appear to bud from the periphery of the saccules (Plate 9A) or from the concave surface of the stack. Other structures commonly seen in the vicinity of the Golgi apparatus include microtubules, centrioles, and striated fibrils (Kojimahara and Kamita, 1986).

The Golgi body is most easily identified when vertically sectioned, but it shows a less characteristic appearance when saccules are tangentially sectioned (Plate 9B), artifactually expanded through poor fixation so as to resemble vacuoles (Plate 9C; see also Plate 19), or almost entirely replaced by vesicles.

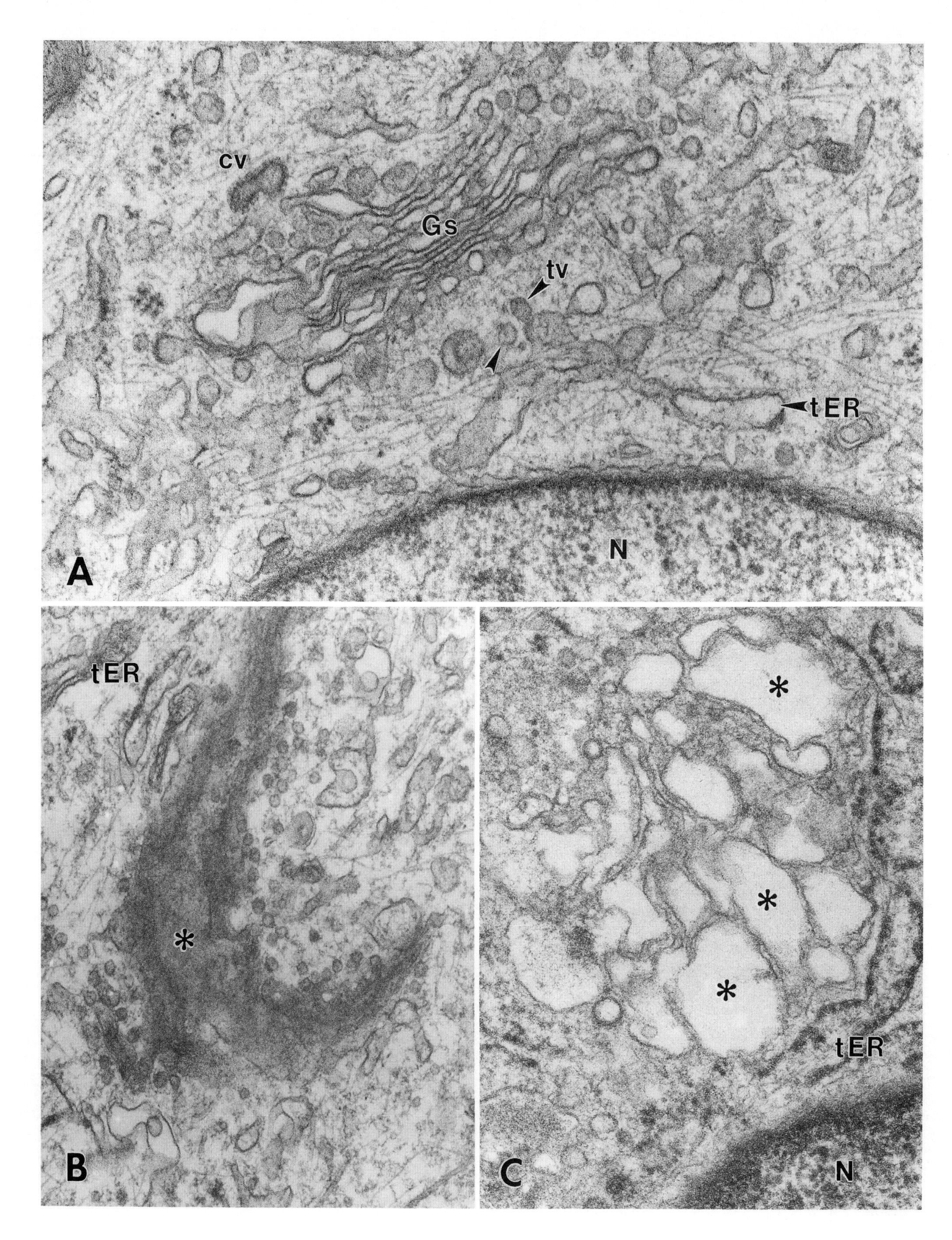
cv
Gs
tv
tER
N
A
tER
*
B
*
*
*
tER
C
N

Plate 9. **A**: A single Golgi apparatus is a typical cytoplasmic location. Note nucleus; transitional endoplasmic reticulum with minimal ribosomal coating away from the Golgi; transitional vesicles; Golgi saccules; "double"-coated vesicle. Gastrointestinal autonomic nerve tumor, abdomen. × 13,100. **B**: Tangential section of Golgi saccule showing an expanse of "gray" membrane (*). Note numerous Golgi vesicles and transitional rER with ribosomes absent from the face nearer the Golgi. Pleomorphic sarcoma, retroperitoneum. × 52,000. **C**: Suboptimally preserved, expanded, and vacuole-like assembly of Golgi saccules (*). Close apposition of several such saccules and the presence of vesicles and transitional rER—all in a paranuclear position—indicate a Golgi body. Uterine leiomyoma. × 63,000. cv, coated vesicles; Gs, Golgi saccule; N, nucleus; tER, transitional rough endoplasmic reticulum; tv, transitional vesicle.

5

Neuroendocrine Granules

Neuroendocrine granules contain peptide or biogenic amine hormones. They are products of the Golgi apparatus, and they mediate endocrine functions on release into the vascular system.

Neuroendocrine granules are found in two broad categories of cells: (1) neuroendocrine cells located in or derived from epithelium and (2) neuronal and related cells. Their tumors include the following:

- *Tumors belonging to the carcinoid–neuroendocrine carcinoma spectrum:* Merkel cell tumors, oat cell carcinomas, gastrointestinal and other carcinoids, medullary thyroid carcinomas, pancreatic islet cell tumors (e.g., insulinomas, glucagonomas), some mucinous carcinomas of breast, pituitary adenomas.
- *Tumors derived from paraneuronal, neuronal, and central nervous system cells*: pheochromocytomas, paragangliomas, neuroblastomas (including olfactory neuroblastoma), peripheral neuroectodermal tumors, gastrointestinal autonomic nerve tumors, central neurocytomas, medulloblastomas, retinoblastomas.

DISTRIBUTION, ULTRASTRUCTURE, AND TERMINOLOGY OF NEUROENDOCRINE GRANULES

Neuroendocrine granules in tumor cells may be nonspecifically distributed in the cytoplasm (Plate 10A) or may be preferentially located in basal areas of cytoplasm, directly under the plasma membrane (Plate 11B) or in cell processes. Some are exocytosed into nonluminal extracellular spaces (so-called **misplaced exocytosis**). They are typically 200–400 nm in diameter, but granules within the wider range of ~80–600 nm are not uncommon; exceptionally they exceed 1 μm. They can survive in tissues embedded in paraffin wax.

Neuroendocrine granules have a single membrane enclosing a matrix (Plate 10C and 10D). Typically, this matrix is very dense, has a very uniform texture, usually has a sharply defined periphery, and is referred to as the **core**. The core is separated from the membrane by a

Plate 10. **A**: Nonspecifically distributed neuroendocrine granules. The large dense granule is a lysosome. Mitochondria are also present. Metastatic carcinoid, liver. ×8500. **B**: Subplasmalemmal neuroendocrine granules (*arrows*). The cells are suboptimally preserved (note the washed out cytoplasmic matrix), but the granules (see **C**) have retained their distinctive morphology. Gastrointestinal autonomic nerve tumor, stomach. ×9700. **C**: Neuroendocrine granules with very dense and homogeneous cores. Note tiny irregularities of granule membrane (*arrowheads*). Same tumor as **B**, ×60,800. **D**: Detailed structure of neuroendocrine granules having diameters of 130–230 nm. Note variation in density of cores, some of which show a granular substructure. Neuroendocrine carcinoma, omentum. ×60,800. ecm, extracellular matrix; ly, lysosome; mi, mitochondria; pm, plasma membrane.

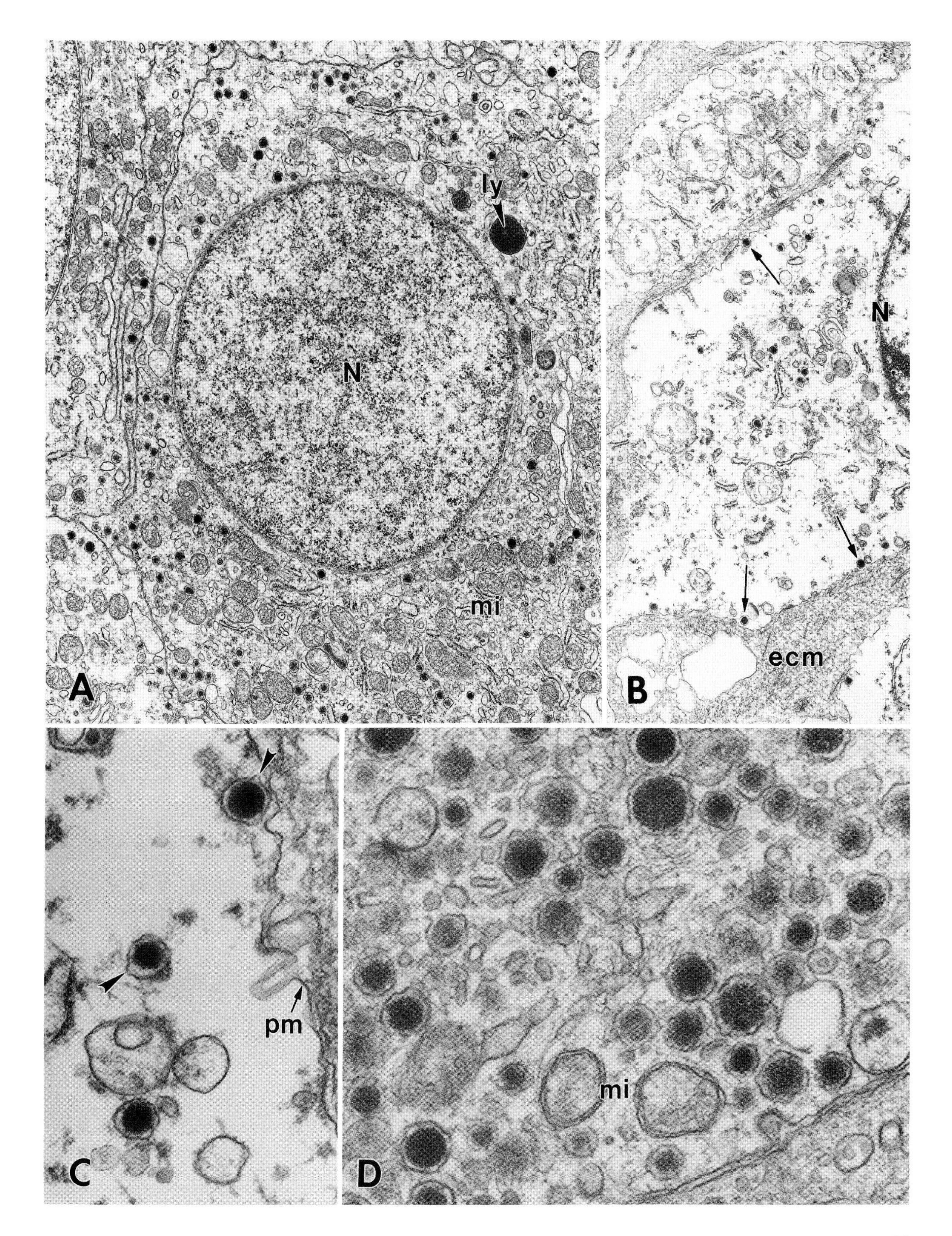
ly
N
mi
A
N
ecm
B
pm
C
mi
D

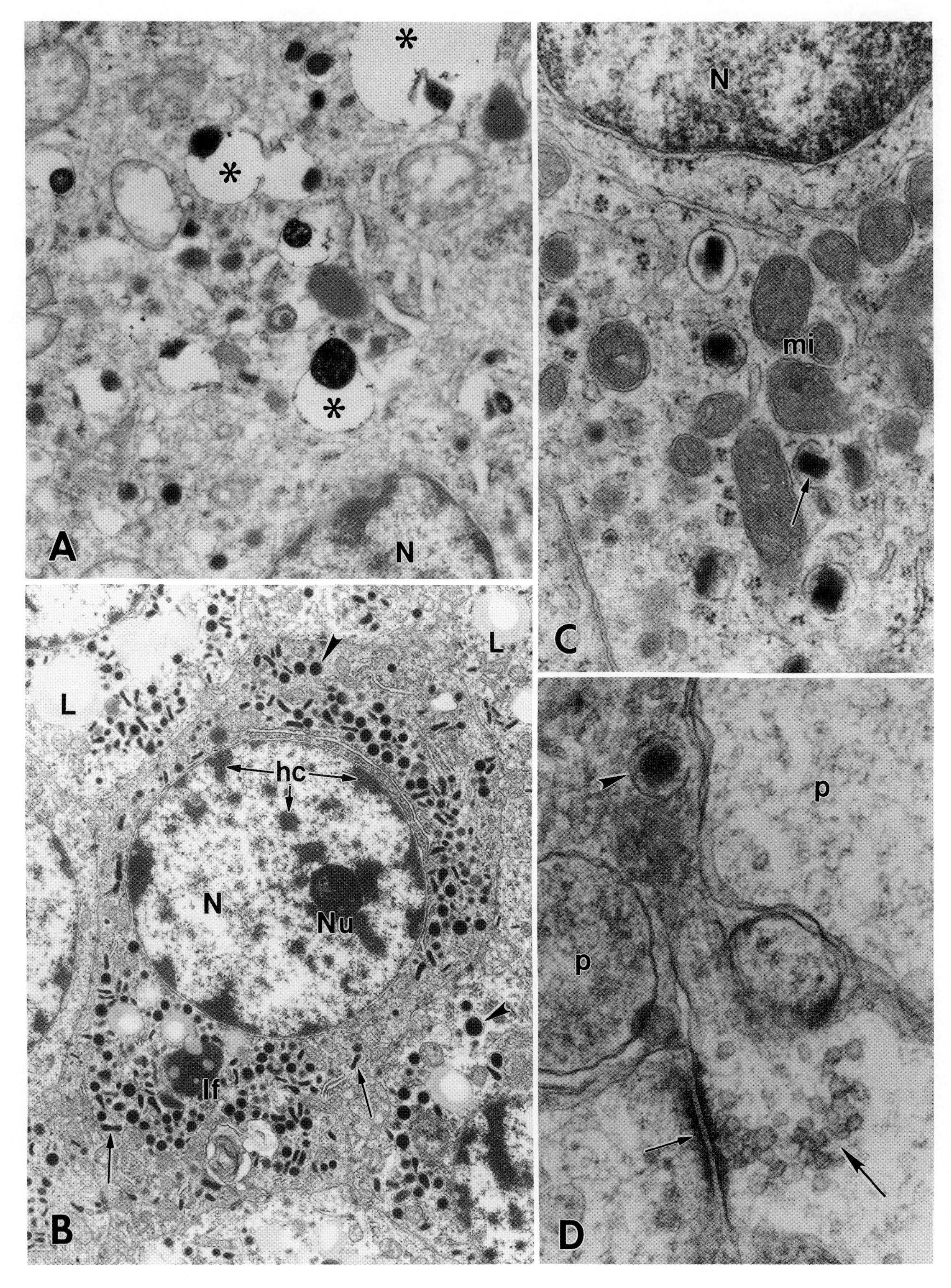
A
N
B
L
L
hc
N
Nu
lf
C
N
mi
D
p
p

clear space or halo, and the membrane often displays slight irregularities (Plate 10C and 10D).

Neuroendocrine granules can be referred to as **neurosecretory**, and they are also described, although less commonly, as **endosecretory**. The term **dense-core granule** is also used, but it should be remembered that this is a purely descriptive designation which can apply just as well to other granules such as primary lysosomes, serous granules, and even some mucigen granules. **Dense-core vesicle** is inappropriate since vesicles are traditionally regarded as having clear contents.

ULTRASTRUCTURAL VARIATIONS ASSOCIATED WITH HORMONAL CONTENT AND ANATOMICAL SITE

Many neuroendocrine granules have a similar appearance with rounded nondescript cores and a size of 200–400 nm in diameter. Some granules, however, have characteristic size ranges or distinctive ultrastructure.

Neuroendocrine Granules with Distinctive Sizes or Appearances in Relation to Specific Lesions or Biochemical Content

- Granules in retinoblastomas, neuroblastomas, medulloblastomas, oat cell carcinomas and certain examples of Merkel cell tumor tend to be small (~80–150 nm diameter).
- Pituitary gland tumors frequently have large granules, reaching up to ~1 μm.
- Norepinephrine granules typical of pheochromocytomas and paragangliomas have eccentric cores (Plate 11A).
- Neuroendocrine granules are biphasic (having rounded and rod-shaped profiles) in many primary neuroendocrine tumors in abdominal and urogenital sites (Plate 11B).
- Insulin granules have crystal-like and occasionally multiple cores (Plate 11C).
- Some glucagon granules have dense foci within the core itself.
- Pointed granules are found in densely granulated growth-hormone-producing pituitary adenomas.

SYNAPTIC VESICLES

The membranous vesicles containing classical neurotransmitter hormone and typically found in neurones are referred to as **synaptic vesicles**, less often as **neuroendocrine vesicles**. In neuronal tumors, they indicate a fairly high level of differentiation. In addition to classical neurotransmitters, synaptic vesicles may contain synaptophysin.

Synaptic vesicles are identified by size (~40–80 nm), appearance (a true membrane enclosing a clear interior), and location (within neuronal processes) (Plate 11D). They are most confidently identified when present in large numbers in the vicinity of **synaptic junctions** (Ojeda et al., 1987; Bertoni-Freddari et al., 1992). One or two vesicles in a neuronal process might be difficult unambiguously to identify as synaptic, rather than, for example, rounded profiles of smooth endoplasmic reticulum which is also found in neuronal cells. Care should also be taken not to misinterpret Golgi vesicles or cross-sectioned microtubules (~25 nm) as synaptic vesicles.

PROBLEMS IN IDENTIFICATION

Neuroendocrine granules are most confidently identified when found in large numbers in tumor cells immunoreactive for hormones and located in tumors with an appropriate clinical and histological setting. In the absence of these supportive data, neuroendocrine granules can be identified by the purely ultrastructural features described above. However, interpretation can be made difficult by poor preservation and by the fact that neuroendocrine granules can resemble other organelles.

Poor preservation can make the texture of the core more coarsely granular or flocculent than usual. It can also make it look rather pale-staining as matrix proteins are leached out; therefore the core may also become much less well defined (Plate 12A). At the same time, the membrane may show exaggerated variations in contour (Plate 12A).

Plate 11. **A**: Eccentric-core norepinephrine granules with large expanded vacuolar areas (*). Pheochromocytoma, adrenal. ×27,200. **B**: Granules of rounded (*arrowheads*) and rod-shaped (*arrows*) profile. Note lipofuscin granule and lipid. Metastatic carcinoid, mesenteric node. × 8400. (Reproduced from Dardick I. *Handbook of Diagnostic Electron Microscopy for Pathologists-in-Training*, New York, Igaku-Shoin, 1996, with permission.) **C**: Insulin granules: Note the angular cores (*arrow*). Normal human pancreas. From a block courtesy of Irving Dardick, M.D. (Toronto). ×38,900. **D**: Cluster of synaptic vesicles (*large arrow*) next to a symmetrical synaptic junction (*small arrow*). Note neuroendocrine granule (*arrowhead*) in adjacent process. Ganglioneuroblastoma, metastatic to cervical node. ×69,000. hc, heterochromatin; l, lipid; lf, lipofuscin granule; mi, mitochondria; N, nucleus; Nu, nucleolus; p, process.

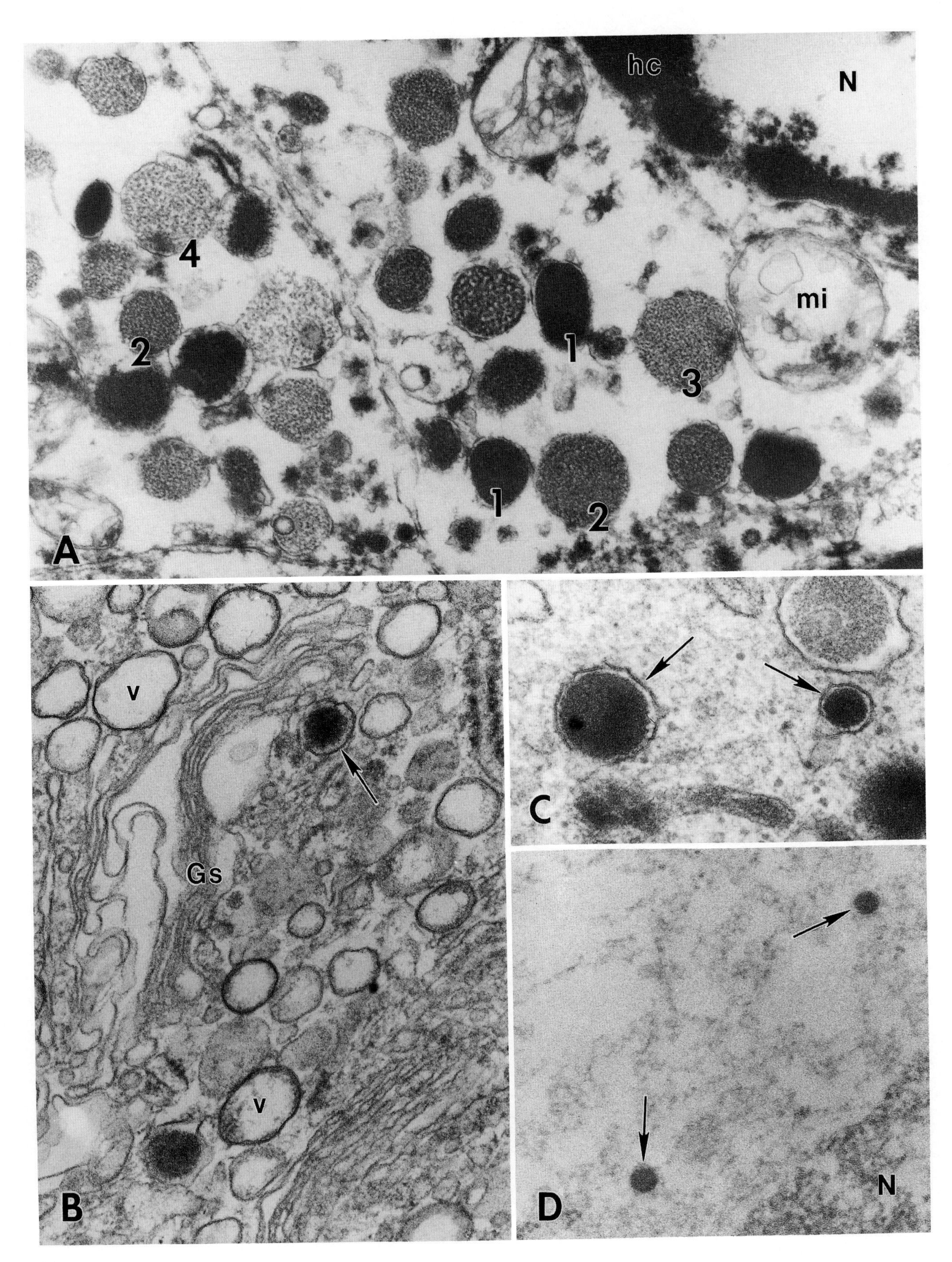
hc
N
4
mi
2
1
3
1
2
A
v
Gs
v
B
C
D
N

Neuroendocrine granules may resemble primary lysosomes (Plate 12B and 12C). Some granules having the rather uniform halo typical of lysosomes (Chapters 6 and 17) have been interpreted as neurosecretory granules in tumors considered as neuronal or neuroendocrine on histological or immunocytochemical grounds. Conversely, some granules *without* typical neuroendocrine features have been shown by immunoelectron microscopy to be truly neuroendocrine. While supportive data from light microscopy will often help in the interpretation of organelles, these observations serve to illustrate the fact that it may be difficult to unambiguously identify small numbers of neuroendocrine granules in nonimmunoreactive tumors without special techniques. These include **immunoelectron microscopy**, the **Grimelius reaction at the EM level** (Vassallo et al., 1971), and the **uranaffin technique** (Plate 12D) (Payne, 1993). The latter uses aqueous uranyl acetate to stain the nucleotide phosphate groups which are found in neuroendocrine granules but which are absent in lysosomes and zymogen, serous, and mucigen granules.

Plate 12. **A**: Variations in substructure and density of matrix and width of halo in a poorly preserved tumor cell (observe washed-out nuclear interior and highly condensed heterochromatin). Note the progressive loss of density (1, 2, 3, 4) and that some of the pale-staining granules (e.g., number 4) almost look like mucigen granules; mucin stains were completely negative in this tumor, however. Primary renal carcinoid. ×38 900. **B** and **C**: Primary lysosomes (*arrows*) resembling neuroendocrine granules in atypical fibroxanthoma (**B**) and in a neutrophil (**C**). **B**, ×44,500; **C**, ×83,300. **D**: Uranaffin-positive granules (arrows) in a peripheral primitive neuroectodermal tumor. ×60,000. Go, Golgi apparatus; Gs, Golgi saccule; hc, heterochromatin; mi, mitochondrion; N, nucleus; v, vesicle.

6

Primary lysosomes

LYSOSOMES: GENERAL PROPERTIES

Lysosomes are organelles which contain acid hydrolases and other degradative enzymes. These mediate digestion at intracellular sites and, when enzymes are secreted, in the extracellular space. Some of these enzymes are cytochemically demonstrable (e.g., acid phosphatase) and are therefore useful in defining lysosomes. Lysosomes are classified functionally into two broad groups. **Primary lysosomes*** contain acid hydrolases but have not yet taken part in digestion; **secondary lysosomes** are organelles in which digestion is proceeding or has taken place.

Intracellular digestion is a virtually ubiquitous process, and one or two lysosomes can be found in almost any cell, but lysosomes are most abundant in the following: granulocytes and macrophages and their proliferative lesions (e.g., granulocytic sarcomas, malignant histiocytosis); true histiocytic lymphomas; follicular thyroid carcinomas (as part of the lysosomal processing of thyroglobulin to thyroid hormones); prostatic adenocarcinomas; the cells in malignant fibrous histiocytoma described as *histiocytic*; neuroendocrine and steroidogenic tumors; Schwann cell and granular cell tumors.

* Primary lysosomes are dealt with here because of the structural similarity that some of them show to neuroendocrine granules. Secondary lysosomes and their organelle precursors associated with endocytosis are dealt with in Chapters 13–17.

ULTRASTRUCTURE OF PRIMARY LYSOSOMES

Primary lysosomes are typically small (~100–300 nm), rounded or oval, and less often elongate or irregular. Since they are Golgi-derived, they all have a single membrane. They have a rather dense, finely and uniformly granular internal matrix; the heterogeneous content associated with digestion is lacking (Chapter 17). Several types of primary lysosome having distinct-

Plate 13. **A**: Eosinophil to show size and distribution of granules. Note crystalline cores (*arrows*). Note also lobated nucleus with large amounts of heterochromatin. Eosinophilic granuloma. ×11,500. **B**: Detail of eosinophil granules showing light and dark crystals (*). Note fine submembrane halo (arrow). Non-neoplastic eosinophil. Granulocytic sarcoma, uterine cervix. ×55,800. **C**: Neutrophil granules with round and rod-shaped profiles. Note solitary glycogen granules in cytoplasmic matrix. Neutrophil in suture granuloma. × 92,300. g, glycogen; hc, heterochromatin; N, nucleus.

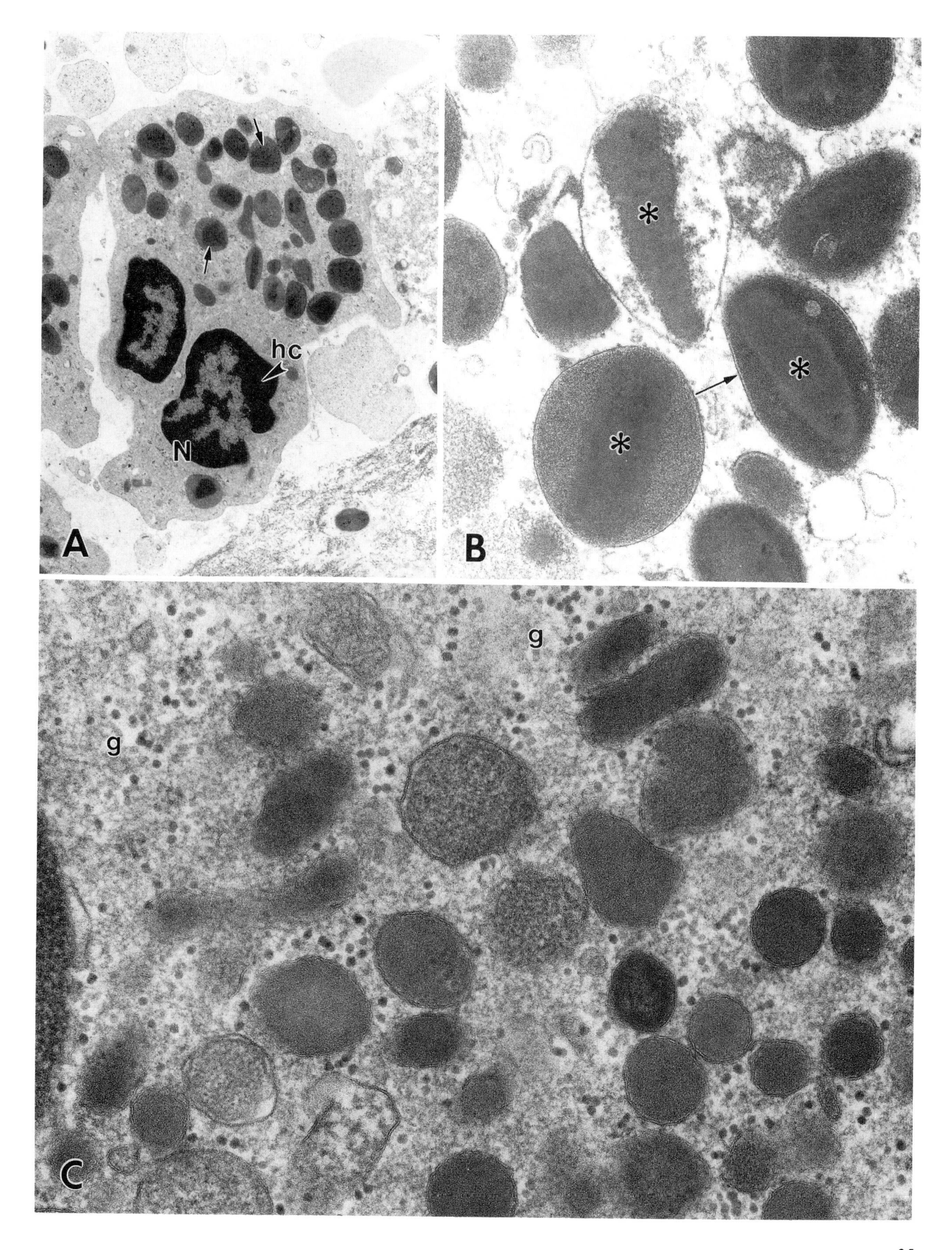

hc
N
A
B
C
g
g

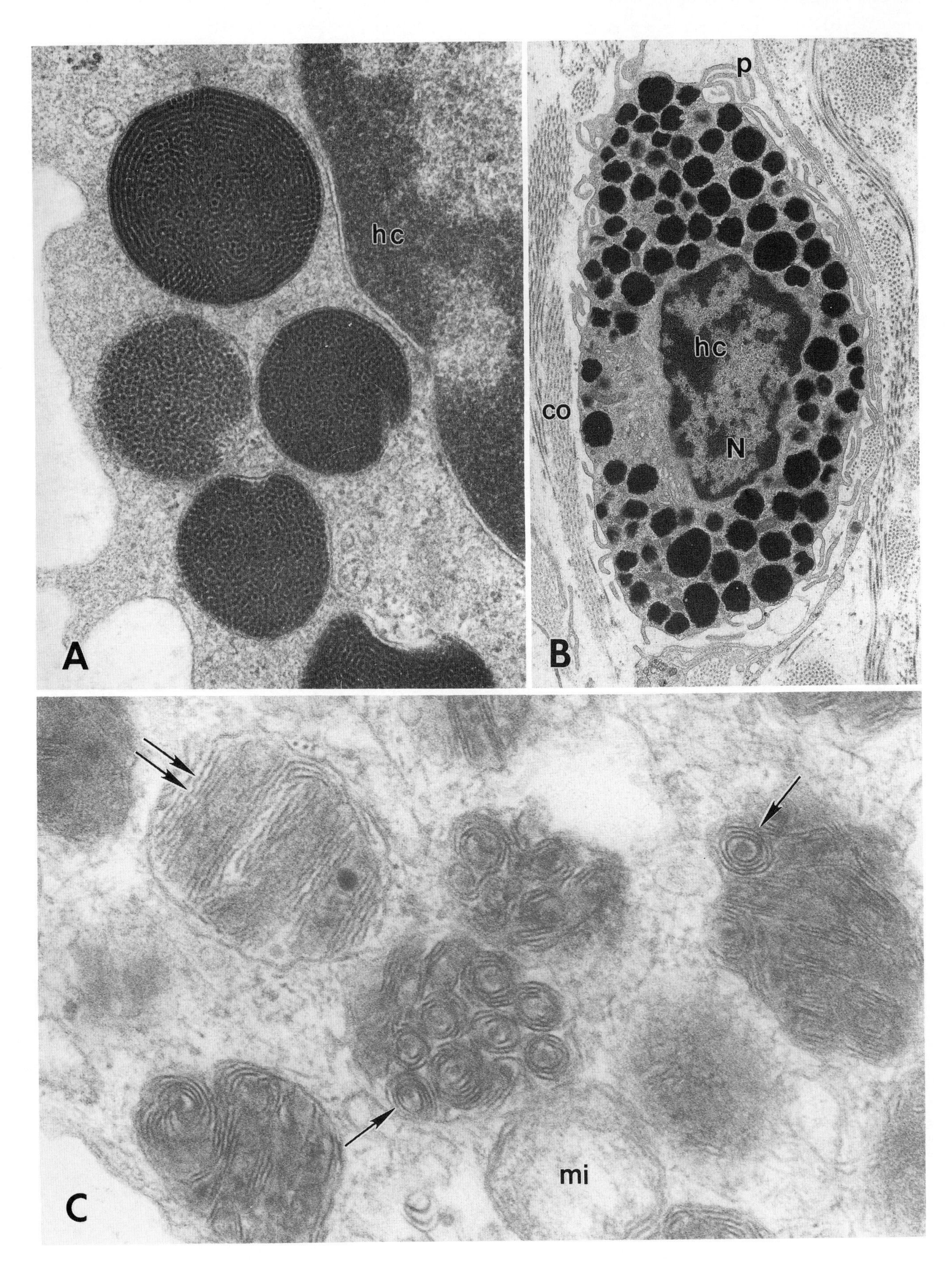
hc
A
p
hc
co
N
B
mi
C

ive internal ultrastructure exist, and some of them are found in and help to define the various classes of granulocyte.

PRIMARY LYSOSOMES IN GRANULOCYTES

Granulocyte primary lysosomes are single-membrane-bound and contain a variably dense internal granular matrix. Typically, they are ~0.5 μm across. **Eosinophil granules** contain a crystalline core in the internal matrix (Plate 13A and 13B); this and the surrounding matrix may be light- or dark-staining depending on the physiological status of the granule and the eosinophil. There is a fine submembranous space of uniform width (Plate 13B), and the matrix, when preserved, is dense and very finely textured.

Mature **neutrophil granules** (Plate 13C) lack internal specialized inclusions and have a uniformly and finely textured granular interior. They may be spherical or rod-shaped and may possess a narrow space beneath the limiting membrane.

Basophil granules contain a matrix consisting of tightly packed coarse dense granules, sometimes arranged in concentric rows (Plate 14A).

Mast cell granules (Plate 14B and 14C) typically have distinctive scroll-like inclusions, but granules in mast cells sampled from human specimens show a wide spectrum of ultrastructure from the characteristic scroll-like inclusions at one extreme to uniform granularity or density at the other (Plate 15). Variable ultrastructural morphology is due partly to variable fixation association with formalin fixation, and to a certain extent to partial degranulation.

PRIMARY LYSOSOMES IN CELLS OTHER THAN MYELOID GRANULOCYTES

Primary lysosomes in reactive cells such as macrophages, plasma cells, and lymphocytes also have a single membrane and usually a dense internal matrix. In macrophages they vary somewhat in size and shape: Although typically small and rounded and ~200–300 nm across, they may also be oval, elongate, or more irregular. Usually, they are mixed with secondary lysosomes which are distinguished mainly by heterogeneous content and varying and larger size (Plate 16A); homogeneous content and small size are suggestive of primary lysosomes.

Primary lysosomes often have a *very* smoothly contoured membrane combined with a submembranous space or halo of very uniform width (Plate 16B and 16C). Some primary lysosomes, however, have the ruffled membrane and variably sized halo reminiscent of neuroendocrine granules (Plate 17A), or no halo at all (Plate 17B). By contrast, the small clear vesicles or vacuoles found in osteoclasts (Plate 17C and 17D) are also primary lysosomes despite their obvious difference from the foregoing examples. Lymphocytes also contain small dense granules; these look dense-cored but are almost certainly primary lysosomes.

Plate 14. **A**: Typical basophil granule showing coarse internal granules, some in concentric rows. Note nucleus with abundant heterochromatin with a coarsely granular substructure. ×57,100. [Micrograph courtesy of U. Schmidt. Reproduced from Schmidt et al. (1988), with permission from Blackwell Science, Ltd.] **B**: Mast cell containing granules which at this magnification appear dense and featureless. Note mast cell processes and collagen fibrils assembled into fibers. Normal intralobular mammary stroma. ×8500. **C**: Mast cell granules in normal myometrium. Note the "scrolls" in cross section (*arrows*) and in long section (*double arrow*). Note mitochondrion. ×95,300. CO, collagen; hc, heterochromatin; mi, mitochondrion; N, nucleus; p, processes.

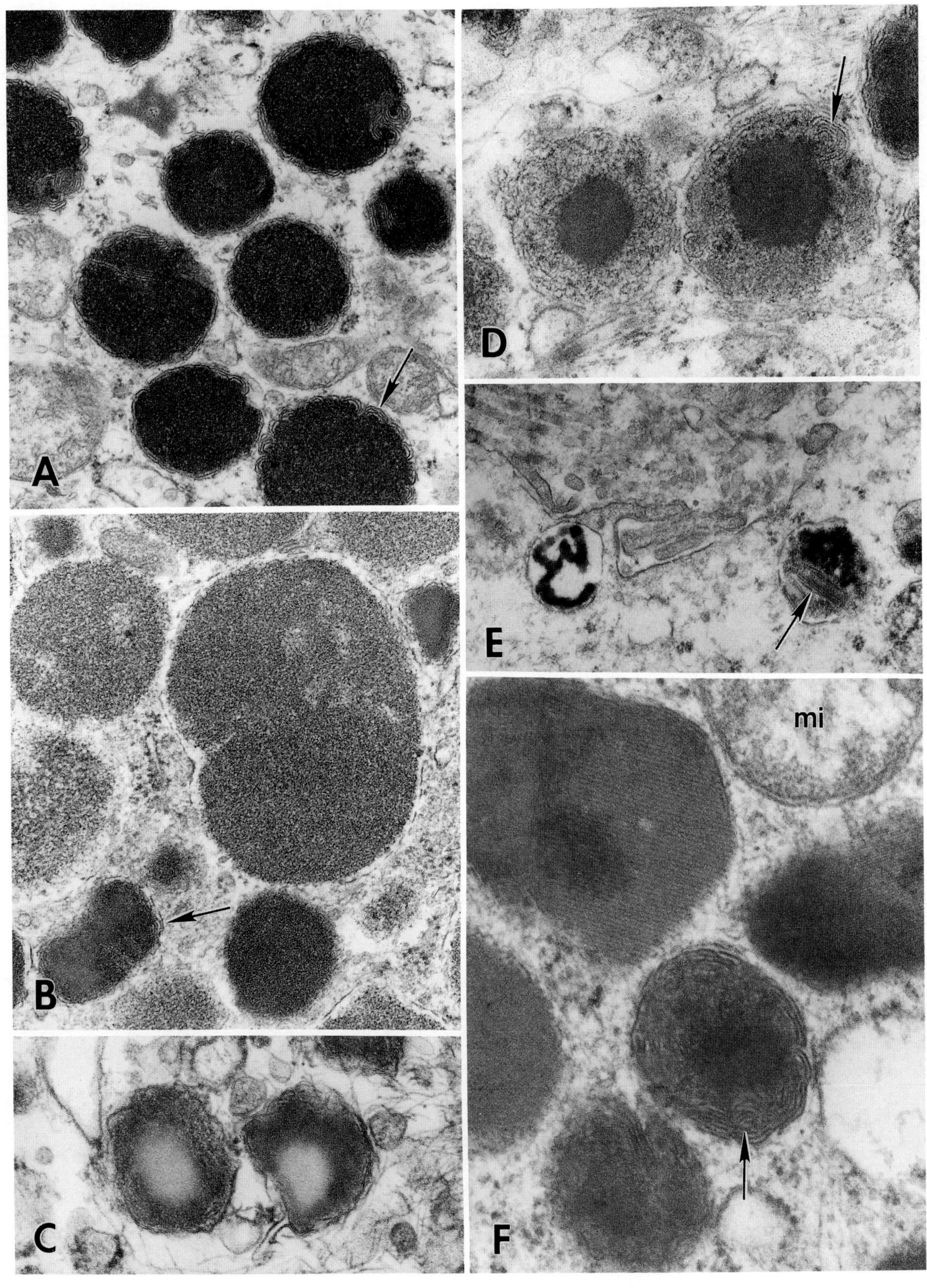
A
B
C
D
E
mi
F

Plate 15. **A–F**: High power views of mast cell granules showing variations in internal structure in various tissues and lesions. The arrows indicate scroll-like structures. **A**: Pleomorphic sarcoma from subcutis of arm. ×39,000. **B**: Paratesticular benign fibroblastic proliferation. ×50,000. **C**: Post-radiation cystitis. ×61,100. **D**: Adult-type rhabdomyoma, larynx. ×60,000. **E**: Diffuse large B-cell lymphoma (multilobated variant), antecubital fossa. ×36,000. **F**: Normal mammary skin. ×80,000. mi, mitochondrion.

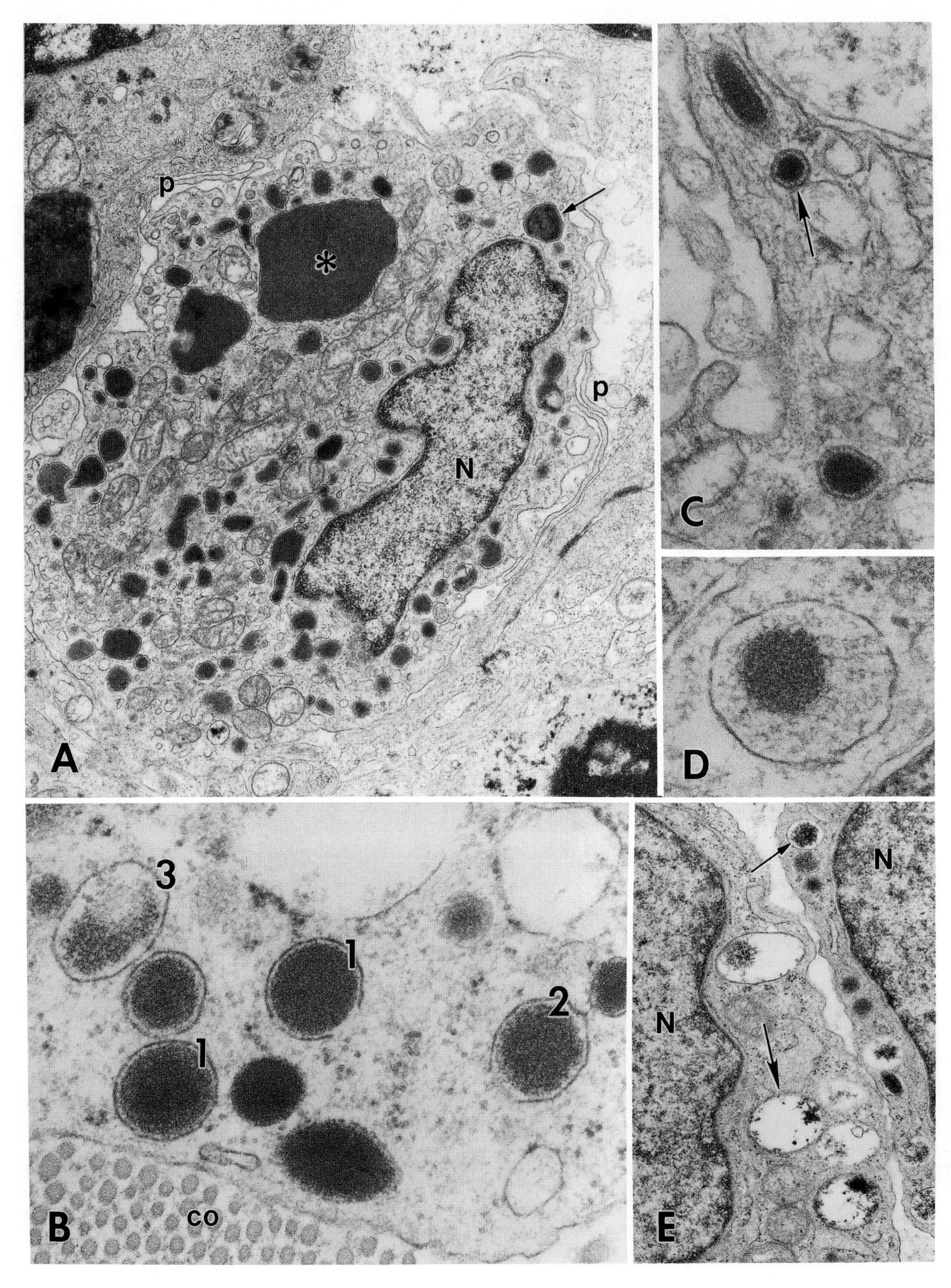
p
*
p
N
A
C
D
3
1
1
2
B
co
N
N
E

Plate 16. **A**: Many small primary lysosomes varying in size and shape. To judge by its heterogeneous content, the body marked by the arrow is a secondary lysosome; the larger body (*) is either a phagososome (possibly containing an erythrocyte fragment) or a secondary lysosome. See also C. Reactive macrophage in a primitive peripheral neuroectodermal tumor, axilla. × 13,000. **C**: Detail of primary lysosomes from **A**. Note how one granule in particular (*arrow*) is reminiscent of a neuroendocrine granule. × 60,100. **B, D, E**: Primary lysosomes of varying morphology from granulocytic sarcoma of uterine cervix. Note that in **B** there is a progression (1, 2, 3) in which the narrow, typically lysosomal submembranous halo is replaced by a more irregular clear space corresponding to an increasingly poorly defined core. Also, the granules in **D** and **E** have reduced or nonexistent/residual cores. Observe collagen in the extracellular space in **B**. **C**, × 62,000; **D**, × 76,000; **E**, × 19,000. CO, collagen; N, nucleus; p, processes.

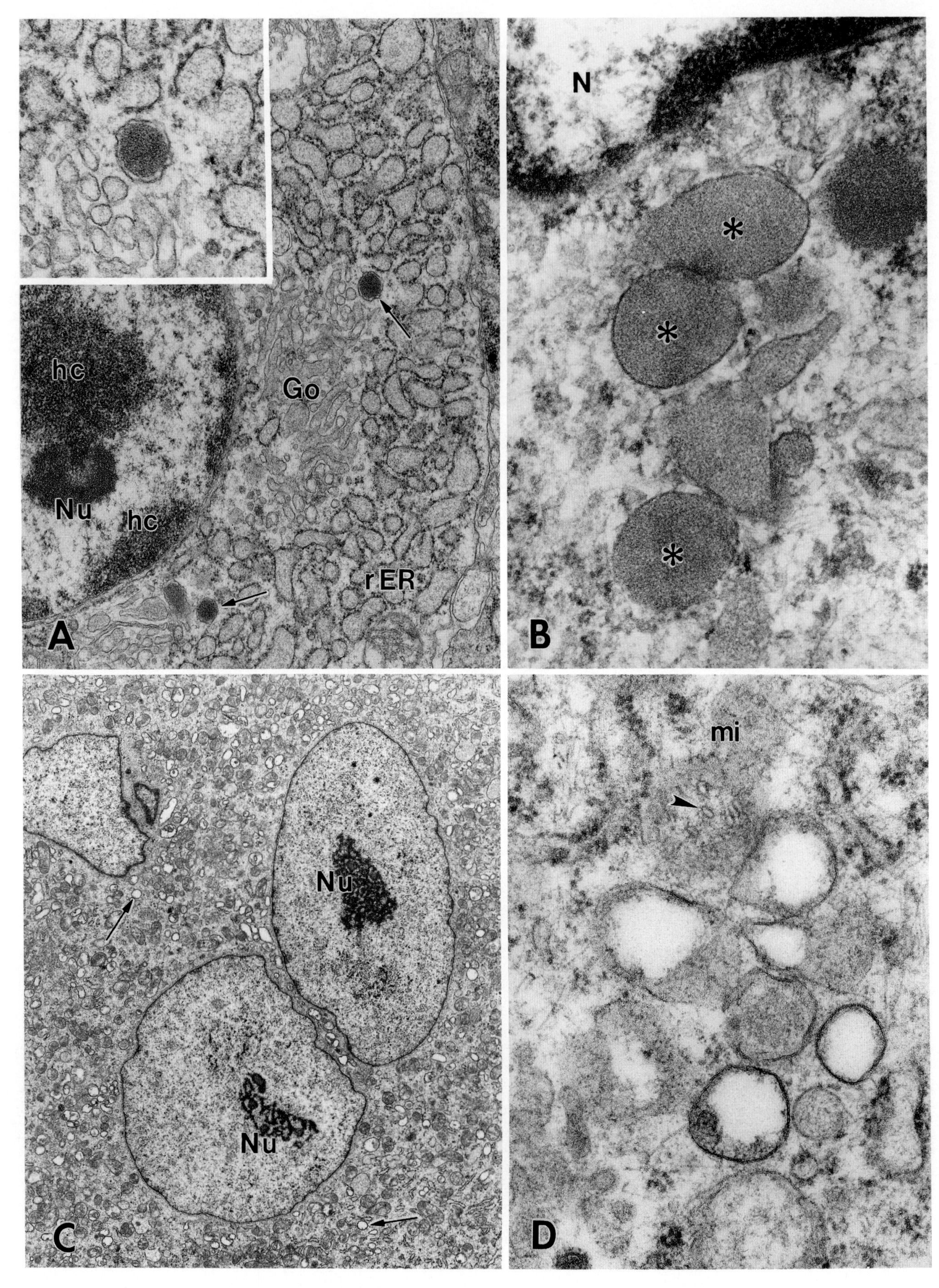
hc
Go
Nu
hc
rER
A
N
*
*
*
B
Nu
Nu
C
mi
D

Plate 17. **A**: Primary lysosomes resembling neuroendocrine granules (*arrows*) in a reactive plasma cell. Note abundant rough endoplasmic reticulum in cytoplasm, along with nucleolus and heterochromatin. Small-cell neuroendocrine carcinoma. ×20,500 (**inset**, ×42,600). **B**: Primary lysosomes without a halo (*) in cervical lymphadenopathy due to reactive monocytoid B cells. ×51,300. **C** and **D**: Reactive multinucleated osteoclastic giant cell in malignant fibrous histiocytoma of giant-cell subtype, breast. Note euchromatinic nucleus and large nucleoli, as well as vesicles (*arrows*). In **D**, note the tangentially sectioned mitochondrion in which cristae are imaged as almost circular profiles (*arrowhead*). **C**, ×4600; **D**, ×60,000. Go, Golgi apparatus; hc, heterochromatin; N, nucleus; Nu, nucleolus; rER, rough endoplasmic reticulum.

7

Mucigen Granules

Mucigen (or **Mucous**) **granules** contain a mixture of glycoprotein and proteoglycan referred to as **mucin**. Once secreted, mucin hydrates and becomes **mucus**, one of the major lubricating agents found over mucosal surfaces.

Mucigen granules are found in mucus-producing mucosal and acinar cells and their neoplastic counterparts as follows: adenomas and adenocarcinomas from respiratory, gastrointestinal, and female reproductive tracts (e.g., mucinous cystadenomas and cystadenocarcinomas of ovary), salivary gland, breast, pancreas; mucoepidermoid carcinomas.

MUCIGEN GRANULE ULTRASTRUCTURE

Typically in normal cells, large numbers of closely packed **mucigen granules** aggregate in the cytoplasm, often excluding other organelles (Plate 18A). They may fuse and form expanses of intracellular mucin. They have a single limiting membrane enclosing a matrix which usually has a flocculent or finely filamentous appearance but which may also be reticulate or homogeneous (Plate 18B). Usually, mucigen granules have a low density. They may also have a small rounded central or eccentric focus of denser material (Plate 18B and 18D), possibly representing protein.

In tumors, mucigen granules can show a range of appearances. Some have a uniformly and much more finely textured content (Plate 18D); some are unusually pale-staining (perhaps poorly preserved); others resemble neuroendocrine granules (e.g., mucigen *progranules*; see Dvorak and Dickersin, 1980). They are interpreted as mucigenic through their occurrence in unambiguous epithelium or adenocarcinoma, sometimes with correlated carbohydrate special stains (e.g., mucin staining, as in Plate 18D).

Mucigen granules can accumulate to produce a signet-ring cell appearance. They can survive paraffin processing.

Plate 18. **A**: Goblet cell almost fully distended with many individually recognizable mucigen granules. Note the lateral cell processes of the adjacent epithelial cells, the lumen, and the dark-staining nucleus. Normal human colon. × 5400. **B**: Mucigen granules with a distinct reticulate substructure (*arrows*). Note small dense cores (*arrowheads*). Nasal mucosa from a child with rhinosinusitis. × 13,300. **C**: Mucigen granules of varying density in a tumor cell profile lacking a nucleus. Adenocarcinoma metastatic to cervical node. × 5300. **D**: Detail of mucigen granules from **C**. Granules are suboptimally preserved; some are completely homogeneous, while others have eccentric granules. × 60,800. lu, lumen; N, nucleus.

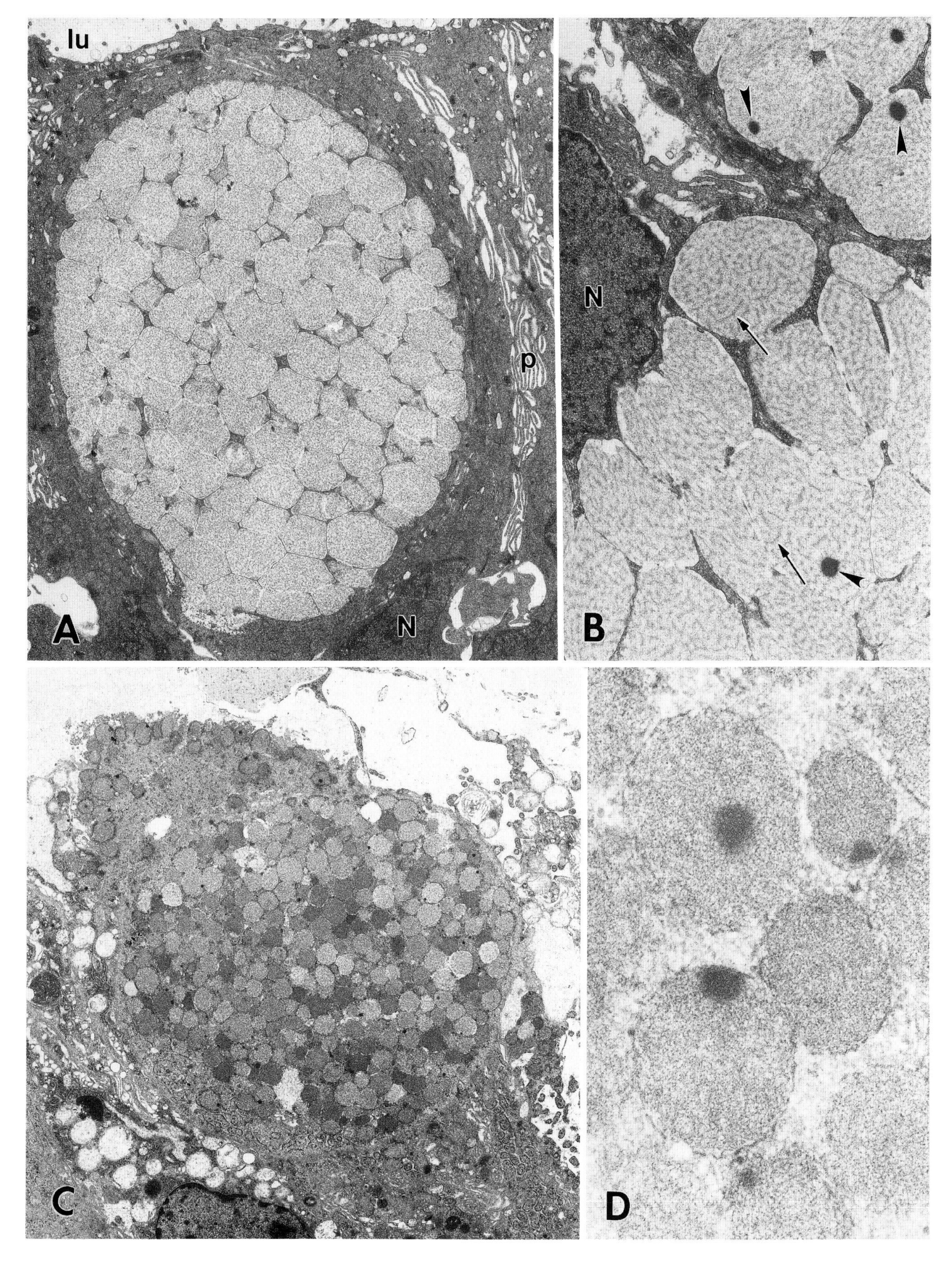
lu
p
N
A
N
B
C
D

8

Serous/Zymogen Granules

Serous and **zymogen granules** of salivary gland and exocrine pancreas contain digestive enzymes or proenzymes which are secreted as part of the saliva or gastrointestinal digestive juices. The terms are commonly used interchangeably, but some authors tend to use **serous** for salivary gland granules and **zymogen** for exocrine pancreas.

In tumors, serous/zymogen granules are found in acinic cell carcinomas, pleomorphic adenomas, solid and cystic pancreatic acinar cell tumors, pancreatic carcinomas and cystadenomas, and pancreatoblastomas.

SEROUS/ZYMOGEN GRANULE ULTRASTRUCTURE

Serous/zymogen granules are up to ~1 μm across. Typically, in normal cells and well-differentiated tumors (Plate 19) they cluster around lumina. They have a single limiting membrane; when well-preserved, the interior matrix has a very high density with no halo (Plate 19, inset).

In pathological conditions, the internal matrix may be unusually pale-staining (Plate 20A), sometimes with a dense and irregular core (Plate 20B). Immature granules, recently detached from the Golgi apparatus, may also have a light matrix, and they may be irregular (Plate 20C). One variant is filamentous (Klimstra *et al.*, 1994). Serous granules can be seen after paraffin embedding.

Plate 19. Epithelial cells containing serous granules (*arrows*). Note abundant rough endoplasmic reticulum, distended Golgi bodies, cellular debris (*) in lumen, and desmosomes (*arrowheads*). Microcystic papillary adenoma, postnasal space. × 7900. **Inset**: Detail of well-preserved zymogen granule showing homogeneous matrix contacting limiting membrane (*arrow*). Normal rat pancreas. × 69,600. lu, lumen.

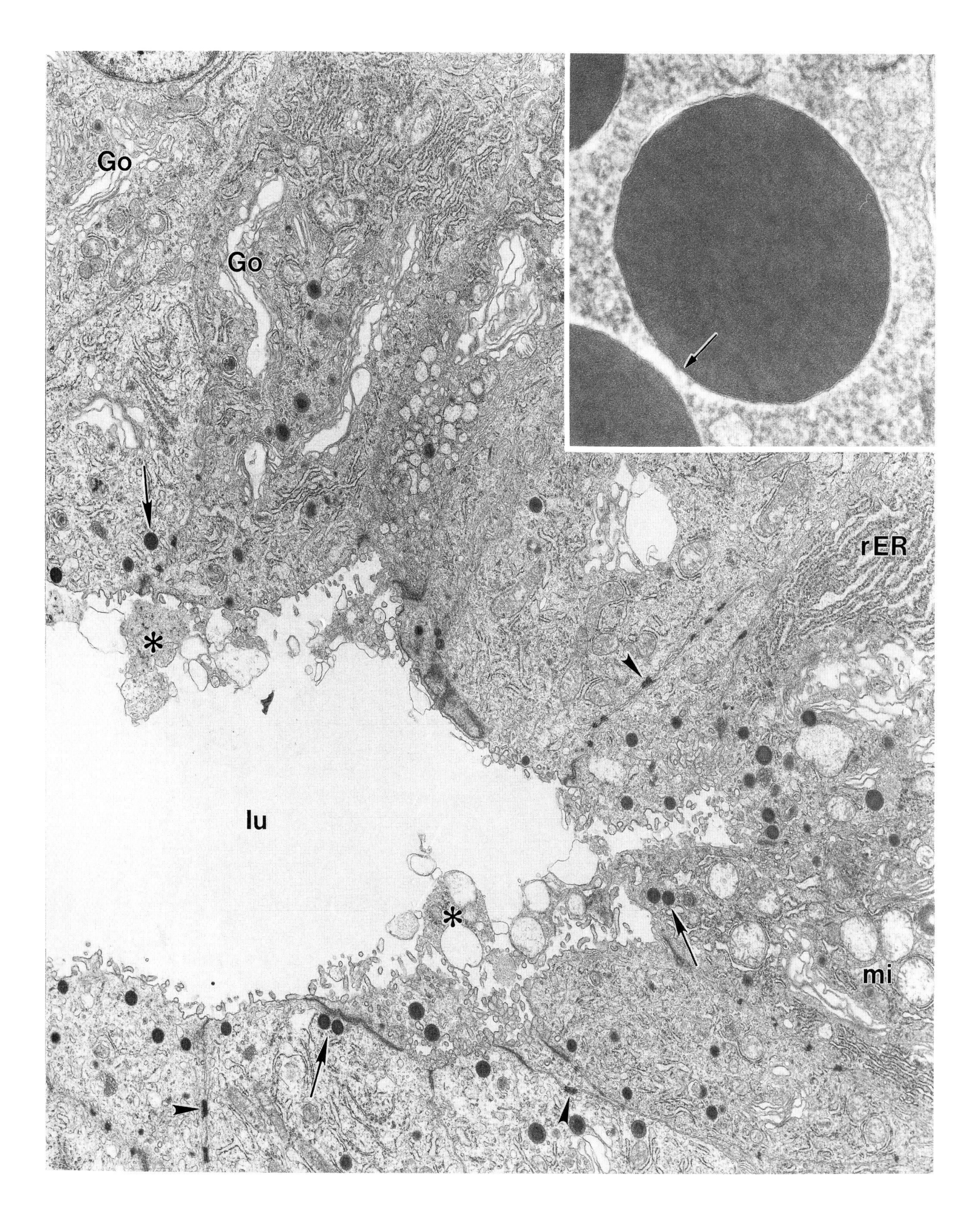
Go
Go
rER
lu
mi

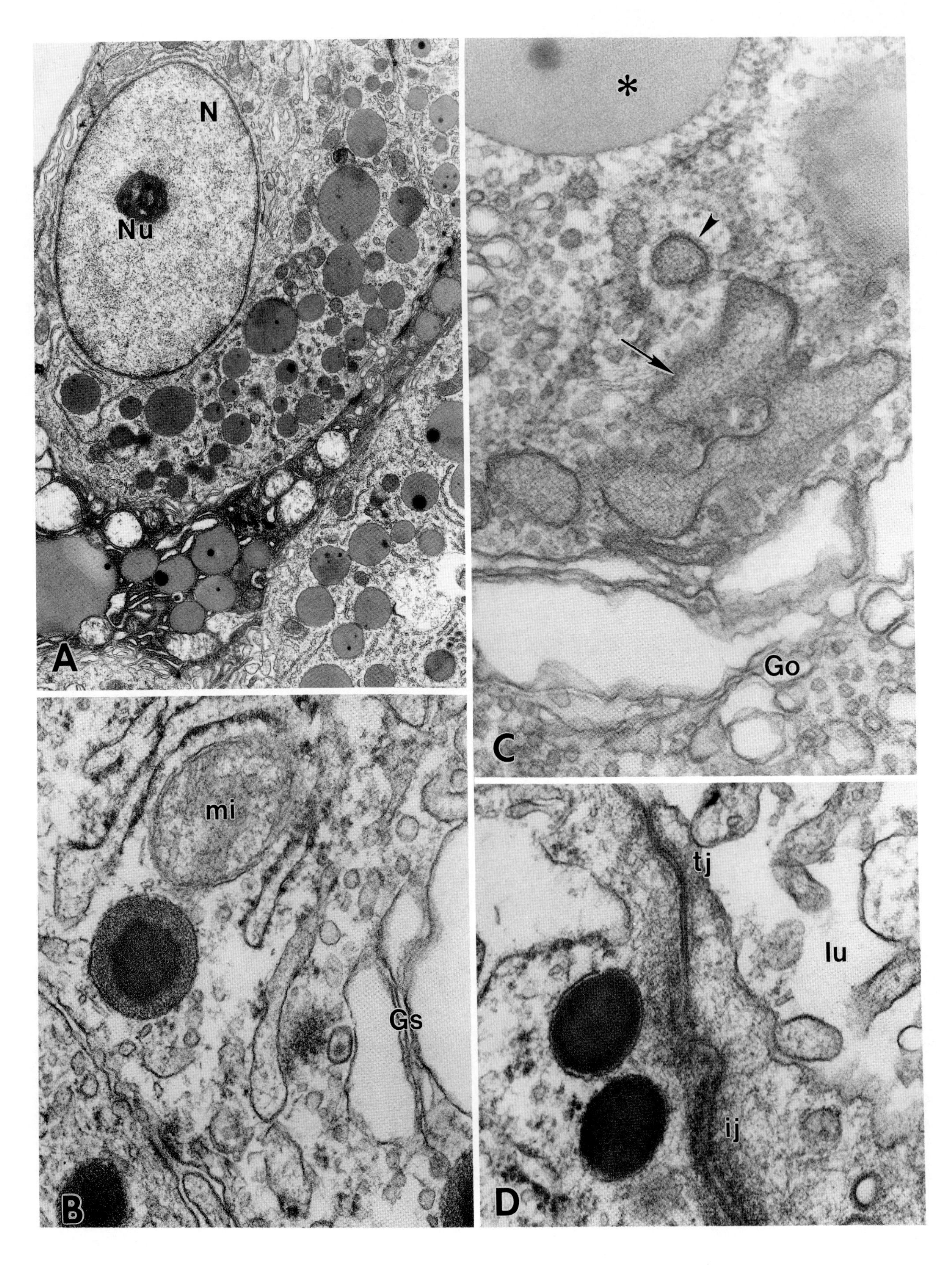
N
Nu
A
*
Go
C
mi
Gs
B
tj
lu
ij
D

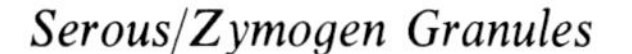

Plate 20. **A**: Light-staining serous granules, also showing small rounded dense granules. Normal human parotid. × 8400. **B**: A small serous granule (350 nm) with a large angular core near artifactually expanded Golgi saccules. Same tumor as **A**. × 60,800. **C**: Immature serous granules—one irregular (*arrow*), one with clathrin coat (*arrowhead*) near expanded Golgi saccules (see Chapter 13 for clathrin). Note homogeneity of the internal matrix of the large mature serous granule (*) and its contact with the limiting membrane. Same tissue as **A**. × 70,700. **D**: Small serous granules near junctional complex (tight junction and intermediate junction) of a lumen; they show a narrow submembranous space and are reminiscent of lysosomes or large neuroendocrine granules. Same tumor as in Plate 19. × 60,800. Go, Golgi bodies; Gs, Golgi saccules; ij, intermediate junction; lu, lumen; mi, mitochondrion; N, nucleus; Nu, nucleolus; tj, tight junction.

9

Miscellaneous Granules

Chapters 5 to 8 describe the most commonly encountered and more easily recognizable granules in human tumor pathology: neuroendocrine granules, primary lysosomes, mucigen granules, and serous granules. These provide evidence of fairly well-defined lines of cell differentiation. This chapter deals with granules which are perhaps less frequently resorted to in tumor pathology, some of which also have a rather complex composition. They include **mammary secretory granules** and granules containing **α-fetoprotein** and **α1-antitrypsin**; they represent some of the largest granules or inclusions encountered in tumor cells.

MAMMARY SECRETORY GRANULES

Several different types of granule are found in mammary epithelium. Granules near lumina are probably **mucigen granules** (Plate 21A and 21B) (Chapter 7); these may have central or eccentric cores (Plate 21A and 21B) or clear contents (Plate 21B).

Mammary epithelium also produces granules containing **lactalbumin**, **casein**, and **gross cystic disease fluid protein**. Distinguishing between these types of exocrine granule can be difficult. The granules in Plate 21C have the large size (1–2 μm) typical of gross cystic disease fluid protein granules; they are also, however, virtually identical in purely ultrastructural terms to serous granules (Plate 19). Establishing the nature of these granules is only possible with clearly defined light microscope immunostaining and preferably immunoelectron microscopy.

Very large granules—several micrometers across—can be found in **mammary carcinomas** (Plate 21D), including those showing **neuroendocrine differentiation**. It is likely that these granules contain either casein or lactalbumin. Again, correlated immunocytochemical

Plate 21. **A**: Lumen with closely applied dense-cored granules with variably sized haloes (*arrows*). In this normal mammary epithelium they are almost certainly exocrine. Normal breast. × 14,800. **B**: Apical cytoplasm showing vesicles and eccentric-core granules close to a lumen (*arrowheads*). The larger dense granules are lysosomes (*arrows*). Normal mammary epithelium. × 35,300. **C**: Large dense granules (*arrows*) near and away from a lumen. Normal breast. × 5400. **D**: Two large granules (*arrows*)—of varying densities, but very homogeneous internal matrix—possibly containing casein. Neuroendocrine granules are also present (*arrowhead*), although their detail is not appreciated at this magnification. Neuroendocrine carcinoma, breast. × 5200. **E**: Putative α-fetoprotein granule (*arrow*). Matrix is not quite completely homogeneous (*arrowhead*). Note mitochondria for size. × 22,000. [Micrograph courtesy of G. Richard Dickersin, M.D. (Massachusetts). Reproduced from Dickersin et al. (1995), with permission from Taylor and Francis.] jc, junctional complex; lu, lumen; mi, mitochondria; N, nucleus; Nu, nucleolus.

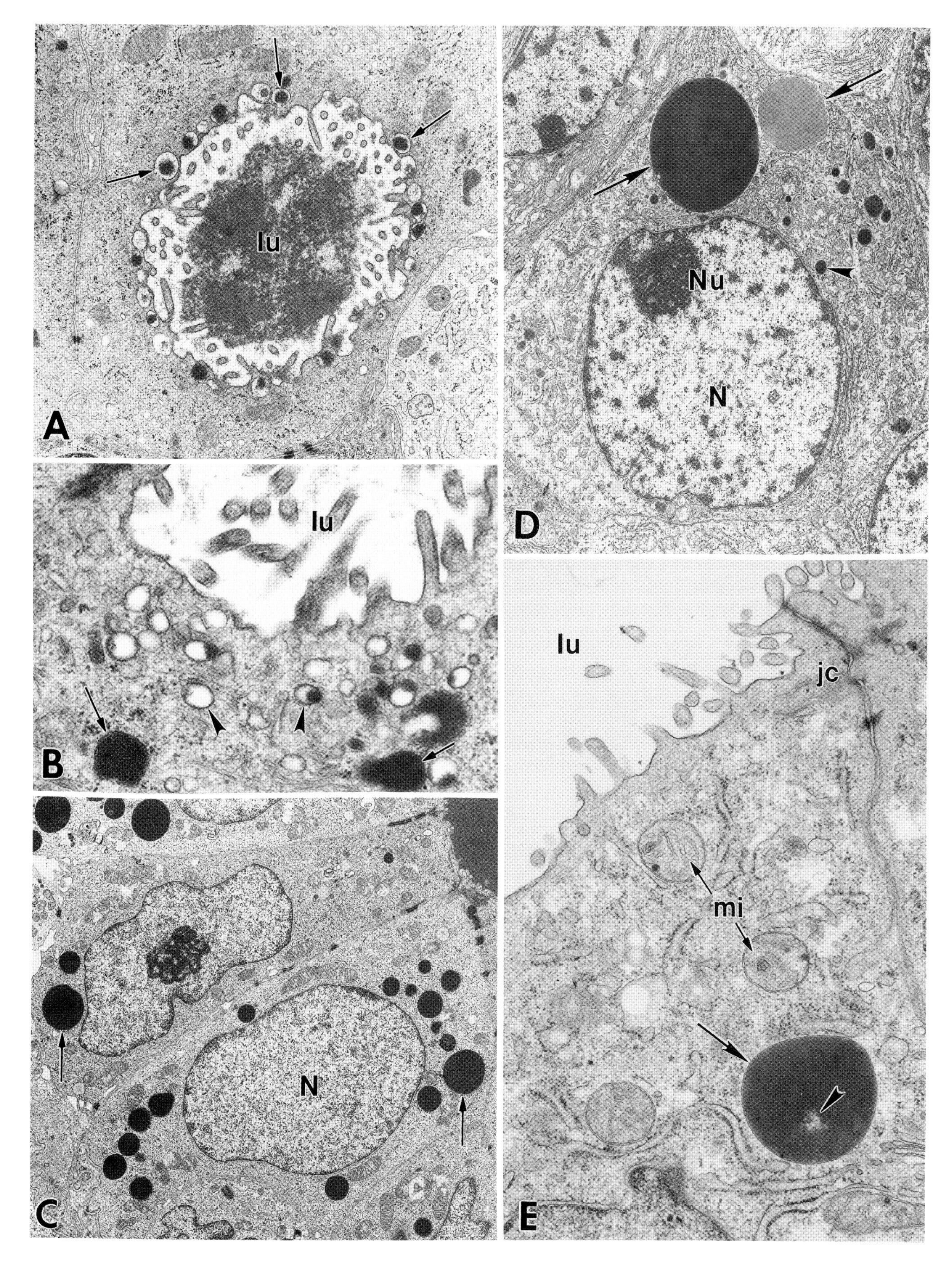
A
lu
B
lu
C
N
D
Nu
N
E
lu
jc
mi

staining or immunoelectron microscopy are needed for unambiguous identification.

α-FETOPROTEIN AND α1-ANTITRYPSIN GRANULES

α-Fetoprotein is produced by fetal liver cells, yolk sac cells, and certain gastrointestinal cells and is used as a marker principally of hepatocellular carcinomas, gonadal germ cell tumors, and yolk sac tumors. On the basis of correlated immunostaining, α-fetoprotein is present in large granules (**electron-dense spheroidal bodies**) having a dense but not always totally homogeneous interior; often one finds small foci of irregularly organized or ill-defined material in the matrix (Plate 21E). α-Fetoprotein granules can often be over 1 μm across (Nakanishi et al., 1982).

Granules of similar size and appearance are found in **embryonal sarcomas of liver**, where by light microscopy they are referred to as **hyaline bodies**. These also are not quite completely homogeneous and there is evidence that, in addition to α-fetoprotein, they contain α1-antitrypsin, ferritin, and albumin; they may therefore be a type of **secondary lysosome** (Abramowsky et al., 1980). Inclusions appearing as intracytoplasmic globules in histological sections have been observed in **pulmonary adenocarcinomas**. They correlate with a complex immunostaining pattern which includes α1-antitrypsin and immunoglobulins and have been described as **secretory glycoprotein** (Scroggs et al., 1989).

10

Melanosomes

Melanosomes are organelles containing tyrosinase and other enzymes for melanin synthesis and are markers of normal and neoplastic melanocytes. They provide the most reliable evidence for melanocytic differentiation, particularly in tumors where the diagnosis is uncertain for reasons of unusual histology or noncontributory immunocytochemistry.

Melanosomes are found in nevi and malignant melanomas (including the soft-tissue variant, clear cell sarcoma), but also in the following tumors not considered primarily melanocytic: melanotic schwannomas, medulloblastomas and neuroectodermal tumors; some carcinoids, paragangliomas, and neuroendocrine carcinomas; Bednar tumor; angiomyolipomas; clear cell tumors of lung.

NORMAL MELANOSOME ULTRASTRUCTURE

Typically, **melanosomes** are rod-shaped or elliptical and are ~200–600 nm long (Plate 22). They exhibit variations in structure due to normal melanization and the neoplastic process. They have a single membrane, and they may show an internal lattice structure, varying quantities of electron-dense melanin, and, occasionally, anomalous structures such as vesicles. Normal melanosomes can be categorized into four types or stages depending on their internal structure and degree of melanization.

Type I melanosomes are the earliest ultrastructurally identifiable melanosomes. They have one or a small number of beaded or periodic filaments in an otherwise clear interior (Plate 22B). They lack melanin. The clear vacuoles found in some malignant melanomas may also be early melanosomes since some have been shown to contain cytochemically demonstrable tyrosinase activity (Hunter et al., 1978).

Type II melanosomes usually have a more elliptical profile. They contain a **lattice inclusion** which may have a variety of patterns: simple cross-striations (Plate 22C), lateral and/or oblique periodicities (Plate 24A), and longitudinal densities (Plate 22D). They lack melanin.

Type III melanosomes differ from type II only by showing varying stages of melanin deposition, sometimes selectively over the longitudinal densities (Plate 22D). **Type IV melanosomes** are fully melanized such that the lattice is obscured (Plate 22A and 22E).

Melanosome lattices and melanin survive paraffin embedding.

Some authors use the term **premelanosome** for type I and type II melanosomes to emphasize that they are *premelanotic* organelles—that is, at a stage *before* melanin formation—and prefer to keep the term *melanosome* for bodies showing a degree of melanization. The terminology using types I to IV, which is recommended here, emphasizes that all of these organelles are developmental variants of a single and continuous maturational sequence.

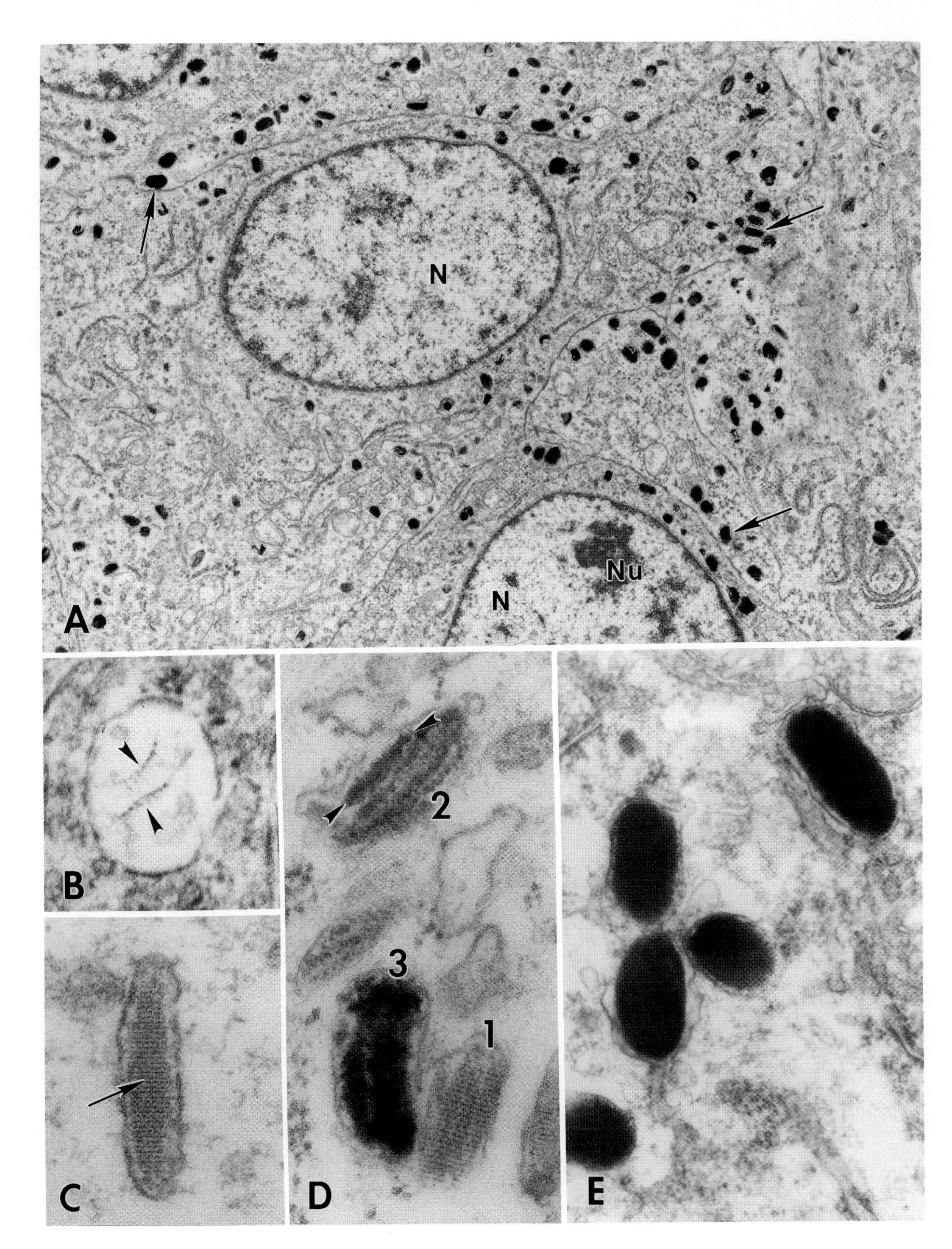
N
N
Nu
A
B
2
3
1
C
D
E

COMPOUND MELANOSOMES

Melanosomes are often found in keratinocytes, macrophages (**melanophages**), and fibroblasts (Plate 23A). Here, they are aggregated within secondary lysosomes and are referred to as **compound melanosomes**. Other less frequently used terms found in the literature include **melanolysosome** and **melanosome complex**. Large compound melanosomes (up to several micrometers across) are equivalent to the **melanin granules** of histological sections. Because compound melanosomes are found in nonmelanocytic cells, melanosomes need to be seen as solitary organelles in the cytoplasm before being considered as evidence of melanocytic differentiation (Plate 22A); the implication is that they have been synthesized by the cell in which they are found.

FINE STRUCTURE OF COMPOUND MELANOSOMES

Compound melanosomes have an enveloping membrane enclosing numbers of melanosomes which are mostly pigmented. Some illustrations in the literature fail to show the limiting membrane very well, possibly through inadequate section staining or membrane preservation (Plate 26F). A diagnostic pitfall here is that the individual melanosomes in the compound melanosome may be misinterpreted as lying free in the cytoplasm and therefore interpreted as a marker of melanocytic differentiation. Although compound melanosomes can be abundant in macrophages, keratinocytes, and fibroblasts, they are also found in malignant melanocytes (Plate 23B).

Since compound melanosomes are lysosomal, they show a spectrum of appearances reflecting different stages of digestion. Some have many intact melanosomes; others show fewer intact melanosomes in a finely granular background matrix, typical of lysosomes and reflecting digestion (Plate 23A). Finally, one may see bodies having almost no recognizable melanosomes; instead, they contain small foci of very dense amorphous pigment, which is almost certainly melanin (Plate 23C).

ABNORMAL NONPIGMENTED MELANOSOMES

Melanosomes in tumors can exhibit abnormalities in internal structure, mainly in the form of aberrant or reduced lattice organization. Melanosomes may contain slightly disordered periodic filaments (Plate 24A) which can be straight, smoothly curved, or angular. It should be remembered that these are filaments in profile only, and three-dimensionally they might be plate-like. Some of the most structurally undifferentiated nonpigmented melanosomes are illustrated in Plate 24B. These lack an obvious periodicity, but some have just-perceptible longitudinal filaments.

RELATIONSHIP OF VACUOLES AND LAMELLA-CONTAINING ORGANELLES TO MELANOSOMES

In some unambiguous melanocytes (from nevi and malignant melanoma) one can see membrane-bound vacuolar structures containing rather loosely arranged lamellae or microvesicles or both (Plate 25A and 25B). Such vacuolar structures have been interpreted as melanosomes (Llombart-Bosch et al., 1988; Ortega et al., 1995). Organelles have also been described which contain larger numbers of lamellae (Plate 25C and 25D). Sometimes, the lamellae are concentrically organized and resemble myelin figures (Chapter 19): these also have been interpreted as melanosomes (Erlandson, 1987; Bhutta, 1993). There is, however, some doubt as to whether these really are melanosomes. Because of their exceptionally sharp delineation in some instances, the trilaminar lamellae may be **lamellar lipid**. There is no compelling evidence that they are analogous

Plate 22. **A**: Tumor cells containing many free cytoplasmic melanosomes, mostly melanized (*arrows*). Malignant melanoma presenting as a black subcutaneous metastatic nodule. ×8500. **B**: Type I melanosome with two periodic filaments (*arrowheads*). Malignant melanoma of soft tissues, foot. ×95,200. **C**: Type II melanosome containing lattice inclusion with a simple transverse periodicity (*arrow*). Metastatic malignant melanoma, axilla ×133,000. Reproduced from Eyden (1989), with permission from Springer-Verlag, and Dardick I. *Handbook of Diagnostic Electron Microscopy for Pathologists-in-Training*, New York, Igaku-Shoin, 1996, with permission. **D**: Several melanosomes. 1, Type II melanosome with a longitudinal aspect to the lattice inclusion; 2, a partially melanized type III melanosome (note the longitudinal pigmentation, *arrowheads*); 3, an almost fully melanized type IV melanosome. Primary malignant melanoma, skin of lower leg. ×112,000. **E**: Intensely pigmented type IV melanosomes. Same tumor as **A**. ×48,400. N, nucleus; Nu, nucleolus.

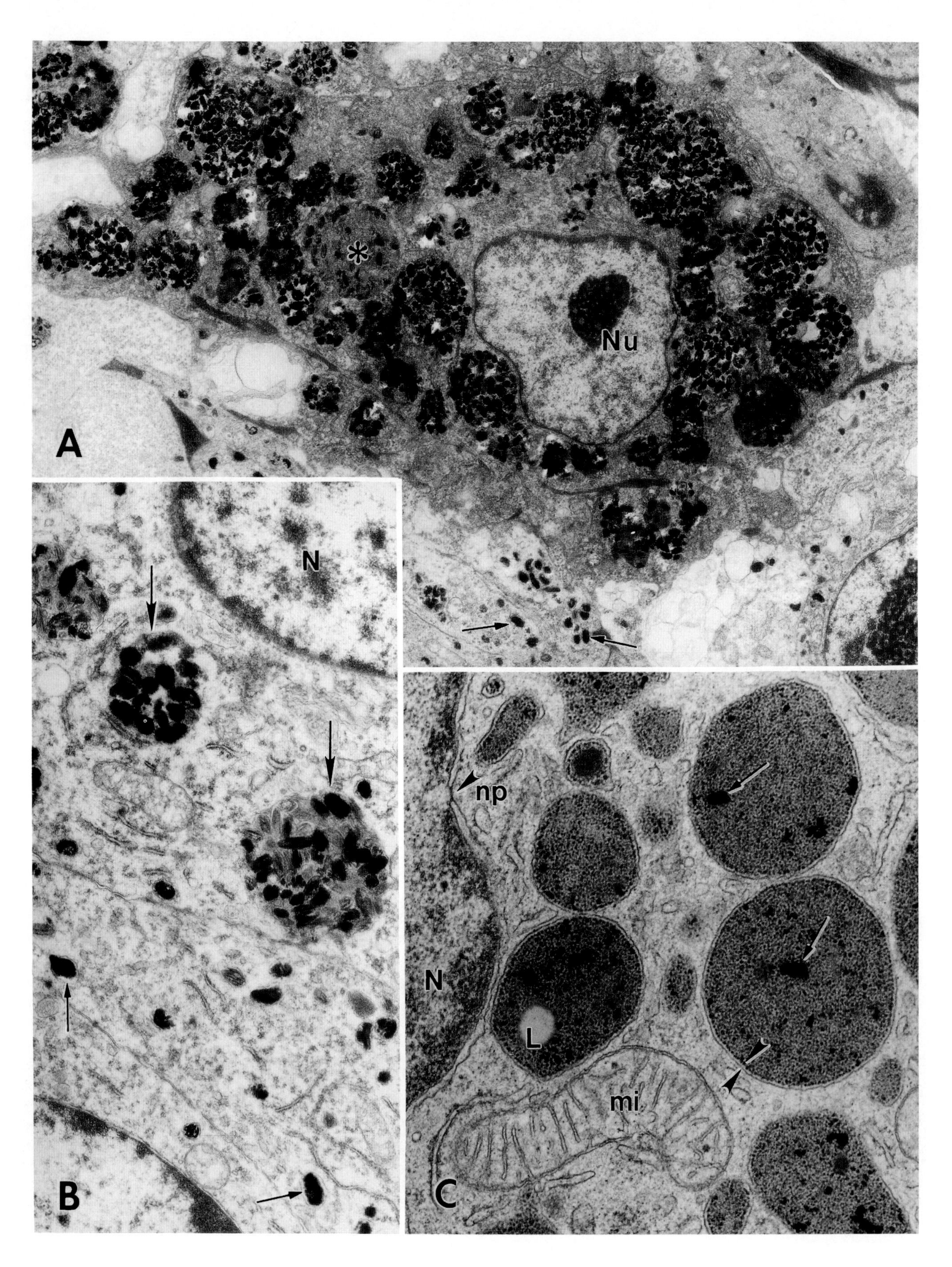
A
Nu
B
N
C
np
N
L
mi

to the solid filaments in melanosomes. In the absence of a clearly defined lattice, the interpretation of a melanosome should depend on seeing *solid filaments* rather than *trilaminar lamellae.*

The clear and mostly structureless vacuoles which are also known in malignant melanoma can be referred to as **vacuolar** or **vesicular melanosomes** provided that tyrosinase can be cytochemically demonstrated within them (Hunter et al., 1978). A further variant with a vacuolar appearance is found in **balloon cell melanoma**; some of these contain a lattice and are therefore type II melanosomes, and it seems that these structures are simply melanosomes artifactually expanded to a vacuole-like appearance (Sondergaard et al., 1980).

ABNORMALITIES OF PIGMENTED MELANOSOMES

Normally, melanin is deposited initially over the longitudinal densities of type II melanosomes (Plate 22D). It can also be laid down in abnormal patterns; in coarse granular clumps over concentric periodic filaments (Plate 26A); over coarse slightly irregularly arranged filaments (Plate 26B); and as amorphous or granular masses within melanosomes where a lattice is absent (Plate 26C). The latter are **granular melanosomes** and are typically rounded or oval. Some organelles exist which lack a lattice and contain just a small amount of dense pigment; because these occur in nevi and malignant melanoma and because the pigment is regarded, reasonably enough, as melanin, the organelles are called **microgranular melanosomes** (Plate 26D and 26E).

One category of exceptionally large rounded melanosome is referred to as **macromelanosomes**. These are as big as compound melanosomes but have a homogeneously dense interior, due to melanin, rather than discretely identifiable melanosomes (Plate 26F).

Macromelanosomes are found in x-linked ocular albinism of the Nettleship–Fall type, *cafe-au-lait* spots in neurofibromatosis, xeroderma pigmentosum, sporadic dysplastic nevi, and spindle cell and epithelioid cell nevi.

Plate 23. **A**: A macrophage with darkly staining cytoplasm full of compound melanosomes. It is surrounded by melanoma cells containing solitary cytoplasmic melanosomes (*arrows*). Asterisk (*) indicates advanced digestion in a compound melanosome. Malignant melanoma presenting as a black subcutaneous metastatic nodule. ×5400. **B**: Malignant melanoma cells containing melanosomes of both free (*small arrows*) and compound (*large arrows*) type. Same tumor as **A**. ×13,300. **C**: Secondary lysosomes in a macrophage containing foci of amorphous pigment (*arrows*) interpreted as remnants of melanosome digestion. Note the background granularity and the fine uniform submembrane halo (*arrowheads*), both of which are typical features of a lysosome. The lipid is also typical of a secondary lysosome. Primary invasive malignant melanoma, interscapular skin. ×36,000. L, lipid; mi, mitochondrion; N, nucleus; np, nuclear pore; Nu, nucleolus.

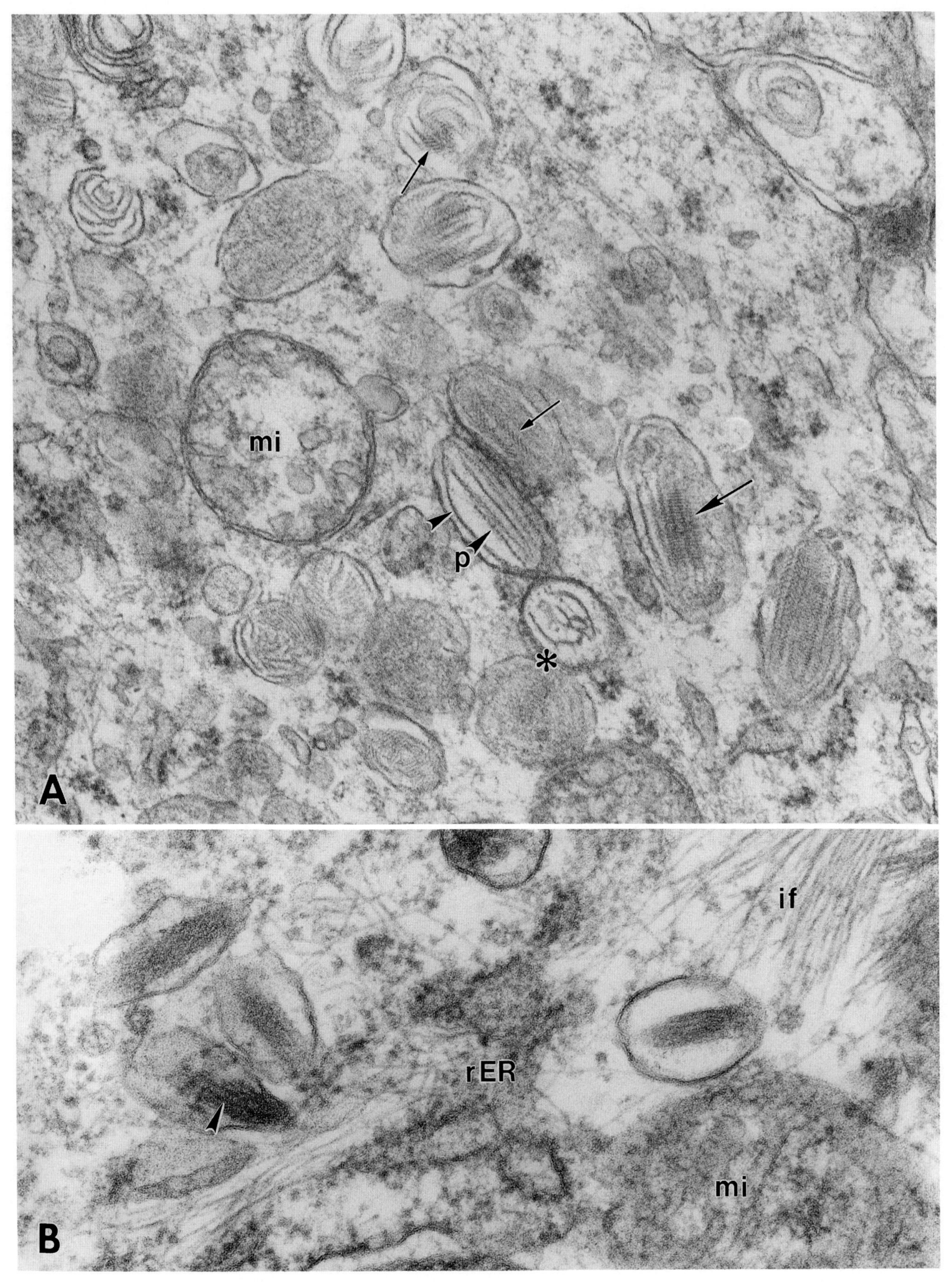
mi
p
A
if
rER
mi
B

Plate 24. **A**: Type II melanosomes. Some have recognizable lattices (*arrows*), some of which show diagonal filamentous periodicity (*small arrows*). In others, the lattice lacks cross-striations but contains filaments, which may be periodic (p) or not (*arrowhead*). Other melanosomes have angular, solid filaments and one or two small granules to indicate a periodic or beaded organization (*). Note the mitochondrion for size. Malignant melanoma, metastatic subcutaneous nodule in groin. ×69,000. **B**: Melanosomes with poorly developed internal structure (*arrowhead*). Intermediate filaments, rough endoplasmic reticulum, and a tangentially sectioned mitochondrion are also present. Malignant melanoma, nasal mucosa. ×95,200. if, intermediate filaments; mi, mitochondrion; rER, rough endoplasmic reticulum.

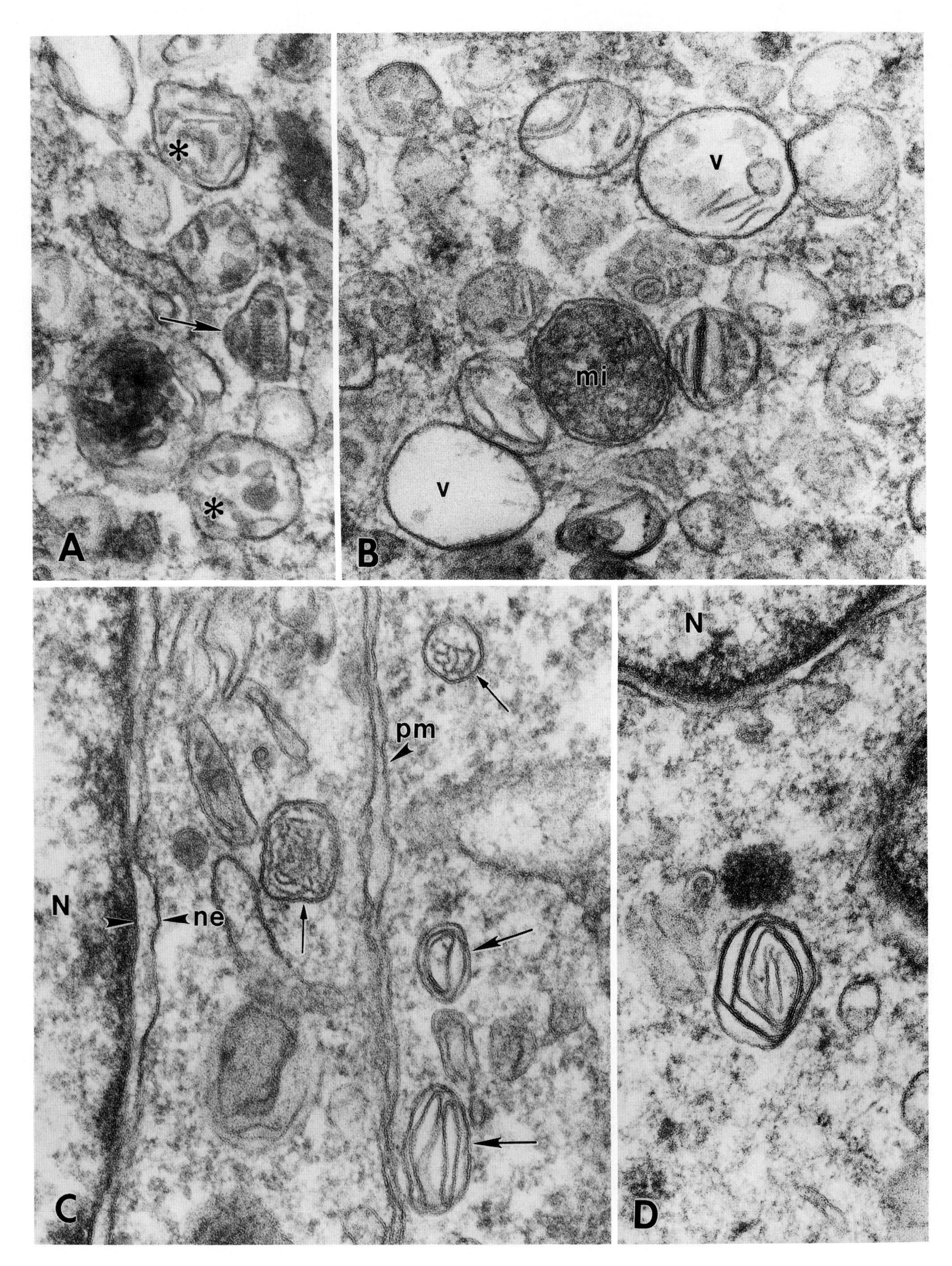
*
*
A
V
mi
V
B
pm
N
ne
C
N
D

Plate 25. **A**: Inconspicuous type II melanosome (*arrow*) and organelles containing vesicles and lamellae (*). **B**: Vacuoles, either with clear contents or containing varying numbers of trilaminar lamellae and vesicles. **C**: Two abnormal melanosomes with reticulate solid filaments (*small arrows*). Note difference from nonmelanosomal bodies containing internal lamellae or membranes (*large arrows*). *In vitro* human malignant melanoma cell line. Specimen courtesy of Dr. Claire Lugassy (Paris). **D**: Organelle with a shape reminiscent of a melanosome, but lacking solid filaments. It contains trilaminar lamellae which are either membranes or, more likely, lamellar lipid. **A**, **B**, and **D** are from an invasive primary vaginal malignant melanoma, positive for S100 protein and HMB45, and containing a small number of unambiguous type II melanosomes (**A**). All figures ×95,200. mi, mitochondrion; N, nucleus; ne, nuclear envelope; pm, plasma membranes; v, vacuole/vesicle.

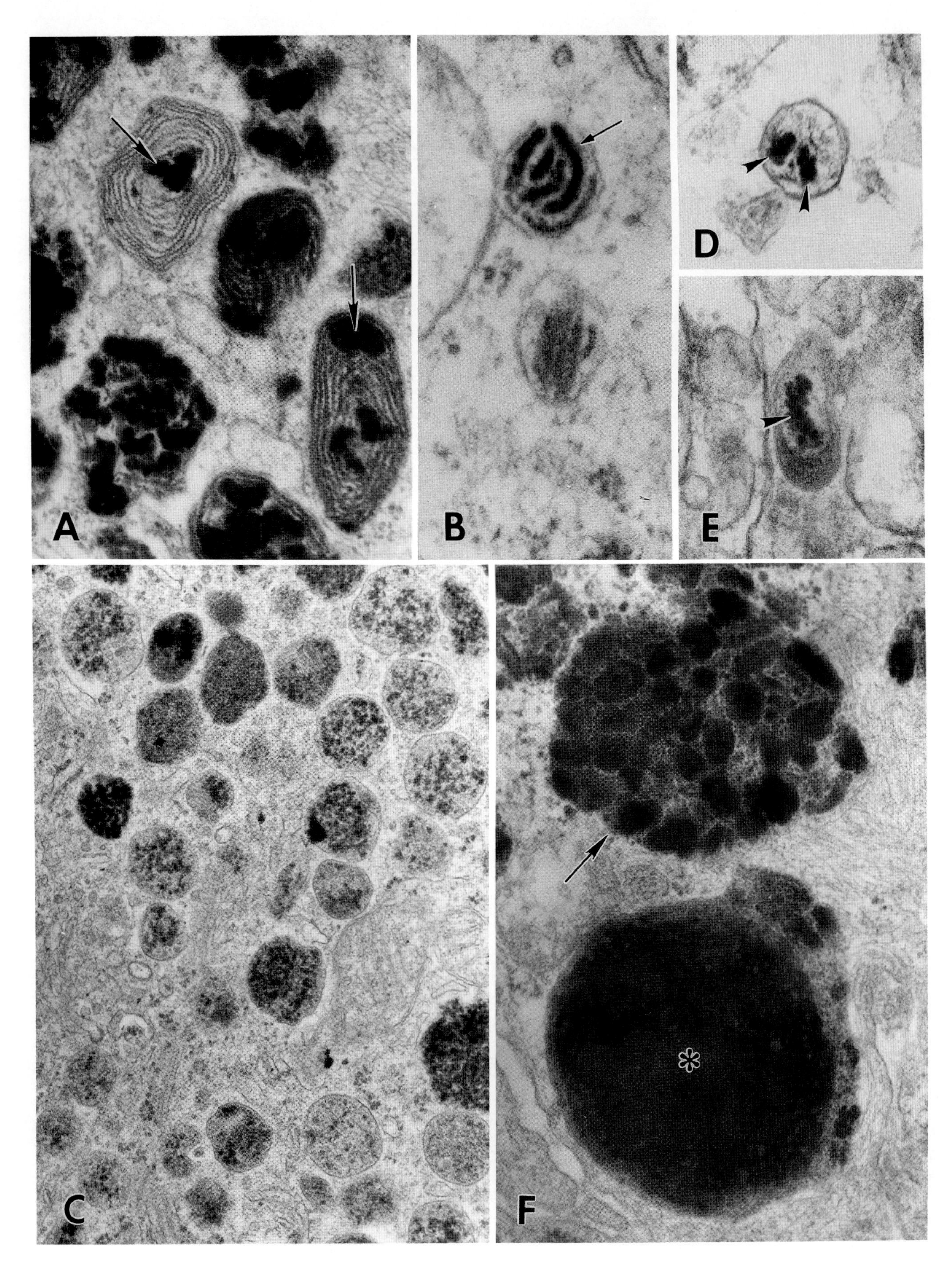
A
B
D
E
C
F

Plate 26. **A**: Abnormal type III melanosomes, some with concentric periodic filaments. Note varying amounts of coarse granules of melanin (*arrows*). Congenital nevocytic nevus. ×53,400. [Micrograph courtesy of Professor Ursula Schnyder and Dr. Barbara Schneider (Zurich). Reproduced from Schneider and Schnyder (1991), with permission from Springer-Verlag.] **B**: Coarse, heavily melanized filaments (*arrow*) in a type III melanosome. Primary malignant melanoma, skin of lower leg. ×112,000. **C**: Granular melanosomes in a malignant melanoma. ×47,000. [Micrograph courtesy of Bruce Mackay, M.D. (Houston, Texas) from a cutaneous metastasis (Mackay and Ayala, 1980). **D** and **E**: Microgranular melanosomes containing highly electron-dense pigment (*arrowheads*), presumed to be melanin. Cellular blue nevus, buttock. **D**, ×70,000; **E**, ×95,200. **F** (*Above*): Typical compound melanosome. Note that at the periphery (*arrow*) the overall limiting membrane is difficult to see. (*Below*) Macromelanosome with homogeneous dense interior (*). Sporadic dysplastic nevus. ×48,000. [Micrograph courtesy of Dr. T. Nakatani (Toyama, Japan). Reproduced from Nakatani and Beitner (1992), with permission.]

11

Weibel–Palade Bodies

Weibel–Palade bodies are organelles specific for vascular endothelium. They contain factor VIII-related antigen, which is released onto the endothelial cell surface where it promotes clotting.

Weibel–Palade bodies are found in angiomatoses, hemangiomas (e.g., capillary and cavernous hemangiomas), neoplastic angioendotheliomatosis, hemangioendotheliomas, and angiosarcomas; in malignant categories of vasoformative tumor, they tend to found in the more differentiated examples or areas.

At low magnification, Weibel–Palade bodies appear as dense cytoplasmic bodies with rounded, oval, or rod-shaped and sometimes elongate profiles (Plate 27A). Typically, they measure about 100 nm in diameter by 1 μm or more in length. Weibel–Palade bodies can resemble other single-membraned organelles with a dense matrix—for example, primary lysosomes, melanosomes, and collagen secretion granules. They differ, however, by their distinctive content, best appreciated at high magnifications.

In cross section, one sees circular profiles (between 5 and 30) which appear to be tubules (Plate 27B). These have a diameter of 20–23 nm, which is similar to that of true cytoplasmic microtubules. There is, however, no evidence that they consist of tubulin protein, and therefore it is preferable to call them **microtubular inclusions**, to indicate their tubular organization and their location *within* the Weibel–Palade body. The term *Weibel–Palade body* is preferred over *rod-shaped microtubulated body*, which is also encountered in the literature.

In longitudinal section, pairs of striations may be seen which represent the walls of the microtubular inclusions (Plate 27C). Sometimes as a result of sectioning geometry or low-magnification photography, the microtubular inclusions are not sharply imaged (Plate 28D). This does not pose an interpretational problem when the organelles are in an obvious vessel, but it may

Plate 27. All micrographs in Plates 27 and 28 are of vessels. **A**: Size and number of Weibel–Palade bodies (*arrows*). Nodular fasciitis in subcutaneous adipose tissue of back. × 13,300. **B**: Cluster of Weibel–Palade bodies in cross section near nucleus (*arrows*). Myxoid sarcoma, hand. × 69,000. **C**: Weibel–Palade bodies in long section, showing microtubular inclusions (*arrowheads*). Note solitary cross-sectioned microtubular inclusion (*arrow*) and pinocytotic vesicles. Spindle cell soft tissue sarcoma NOS, foot. × 80,100. **D**: Two light-staining and immature Weibel–Palade bodies (*large arrows*), one showing clathrin coat (*small arrow*) (see Chapter 13 for clathrin). Arrowheads indicate microtubular inclusions. Primary fibrolamellar hepatocellular carcinoma. × 95,200. (**B** and **C** reproduced from Dardick I. *Handbook of Diagnostic Electron Microscopy for Pathologists-in-Training*, New York, Igaku-Shoin, 1996, with permission.) ecm, extracellular matrix; if, intermediate filaments; lu, lumen; N, nucleus; pv, pinocytotic vesicles.

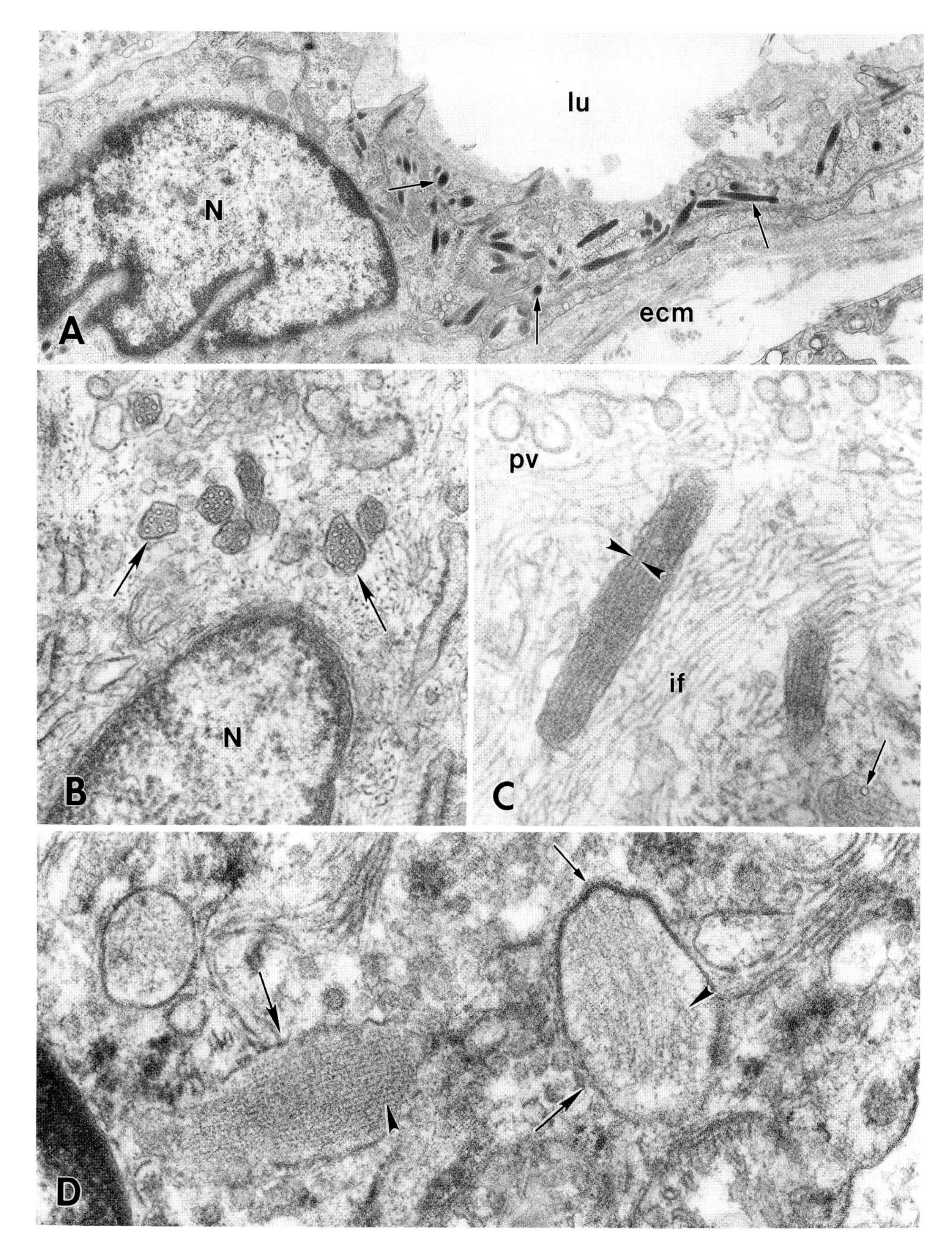
lu
N
ecm
A
N
B
pv
if
C
D

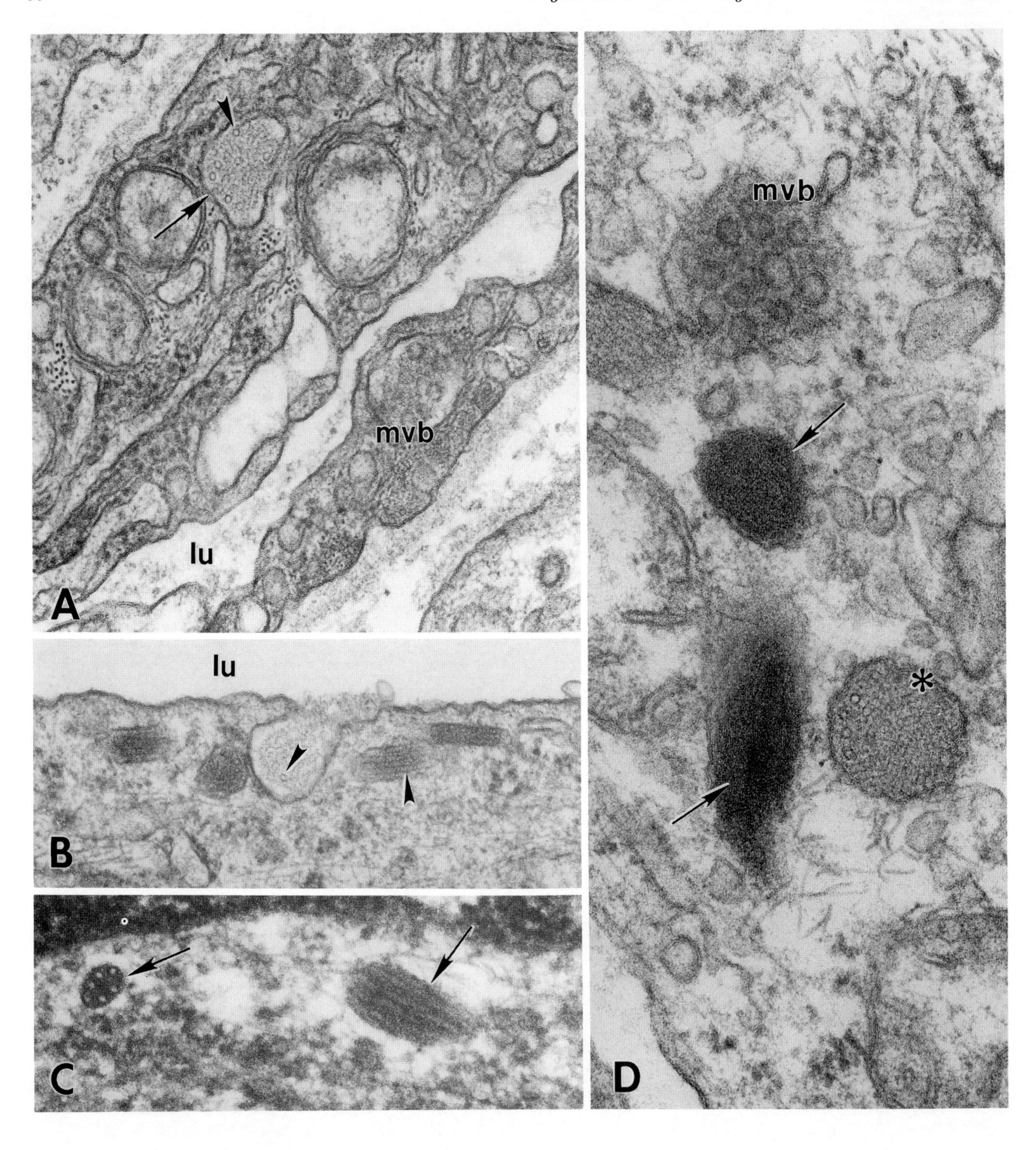
mvb
lu
A
lu
B
C
mvb
*
D

in a diagnostically problematical and putative vascular tumor. It is therefore highly desirable to see sharply defined microtubular inclusion profiles in any organelle being considered as a Weibel—Palade body (Kitagawa et al., 1985). In addition, some bodies with a crystalline or laminated content have been considered as Weibel–Palade bodies; without demonstrating microtubular inclusions or ultrastructural immunolabeling for Factor VIII-related antigen, however, the interpretation of these organelles as Weibel–Palade bodies must be considered tentative (Fujioka et al., 1995).

Although Weibel–Palade bodies are most often rod-shaped and electron-dense, they may also be more rounded and light-staining (Plates 27D and 28A): these tend to be larger (typically more than 200 nm) than the denser, rod-shaped examples. They are light-staining when they are immature and have just developed from the Golgi apparatus (Plate 27) and also when they are about to discharge their contents into the lumen (Plate 28B); here, the microtubular elements may be scarce.

Weibel—Palade bodies should not be confused with multivesicular bodies which are also found in endothelium and which also have a single membrane enclosing circular profiles. Multivesicular bodies are larger (typically 200–800 nm across), and the internal circular profiles are of true membrane (they represent microvesicles); they measure at least 60 nm in diameter and are, therefore, several times bigger than the microtubular inclusions in Weibel–Palade bodies (Plate 28D).

Weibel–Palade bodies can survive paraffin embedding (Plate 28C).

Plate 28. **A**: Pale-staining Weibel–Palade body (*arrow*) containing microtubular inclusions (*arrowhead*). Note multivesicular body and lumen. Alveolar soft-part sarcoma, thigh. ×69,000. **B**: Weibel–Palade body possibly discharging contents which include a single microtubular inclusion (*arrowhead*) and surrounded by more typically rod-shaped and denser Weibel–Palade bodies, also showing microtubular inclusions (*arrowhead*). Pseudoangiosarcomatous squamous cell carcinoma, breast. ×69.000. [**A** and **B** reproduced from Eyden (1993), with permission from the *Journal of Submicroscopic Cytology and Pathology*.] **C**: Weibel–Palade bodies (*arrow*) from a deparaffinized block. Metastatic malignant melanoma to subcutis. ×104,000. **D**: Tangentially sectioned Weibel–Palade bodies failing to show internal microtubular inclusions because of sectioning geometry (*arrows*), and rounded pale-staining Weibel–Palade body (*) illustrating distinction from multivesicular body. Epithelial mesothelioma, peritoneum. ×98,000. [Reproduced from Eyden (1989), with permission from Springer-Verlag.] lu, lumen; mvb, multivesicular body.

12

Collagen Secretion Granules: Intracellular Collagen

COLLAGEN SECRETION GRANULES

Collagen secretion granules are Golgi-derived organelles which contain and transport to the cell surface the collagen precursor, **procollagen**. Therefore, they are markers of all cells actively producing collagen precursor for secretion.

Collagen secretion granules are found in all matrix-forming cells—fibroblasts, myofibroblasts, odontoblasts, osteoblasts, chondroblasts and corneal epithelium; in tumor cells they have been seen in myofibrosarcomas, myofibromatoses, a spindle cell carcinoma, and malignant fibrous histiocytomas (Plate 29).

ULTRASTRUCTURE OF COLLAGEN SECRETION GRANULES

In low-magnification micrographs, collagen secretion granules appear as dense rod-shaped bodies in the Golgi zone near the nucleus (Plate 29A). Immature granules are more rounded and more pale-staining and are full of fine procollagen filaments (Plate 29B). With maturation, the filaments aggregate to produce bundles, and the organelle becomes more rod-shaped (Plate 29C–F). Foci of clathrin (Chapter 13) indicate pinching-off of limiting membrane as the granule becomes less voluminous (Plates 29E and 30A). Later, the filamen-

Plate 29. **A**: Collagen secretion granules (*arrows*) in Golgi zone and peripheral cytoplasm. Note larger mitochondria. Myofibroblast from fibroma of tendon sheath. ×12,900. **B**: Very immature collagen secretion granule (*arrow*) containing finely filamentous procollagen and probably having just developed from a Golgi saccule. Myofibrosarcoma, dermis of temple. ×70,700. **C**: Collagen secretion granules (*arrows*) showing aggregation of procollagen filaments to form a single bundle. Note the hint of dense nodes (*arrowheads*). Same tumor as **B**. ×69,000. [**B** and **C** reproduced from Eyden et al. (1991), with permission from Taylor and Francis.] **D**: Collagen secretion granule with two procollagen filament bundles (*arrows*). Spindle cell carcinoma, larynx. ×69,000. **E**: Collagen secretion granule with procollagen filament aggregate (*arrow*) and clathrin-coated vesicle probably budding from membrane (*arrowhead*). Human granulation tissue myofibroblast. ×98,000. **F**: Procollagen filaments may collapse under conditions of poor fixation and may not be individually resolvable. The limiting membrane also frequently has an irregular outline. Giant cell malignant fibrous histiocytoma of breast. ×69,000. ecm, extracellular matrix; Gs, Golgi saccules; mi, mitochondrion; rER, rough endoplasmic reticulum.

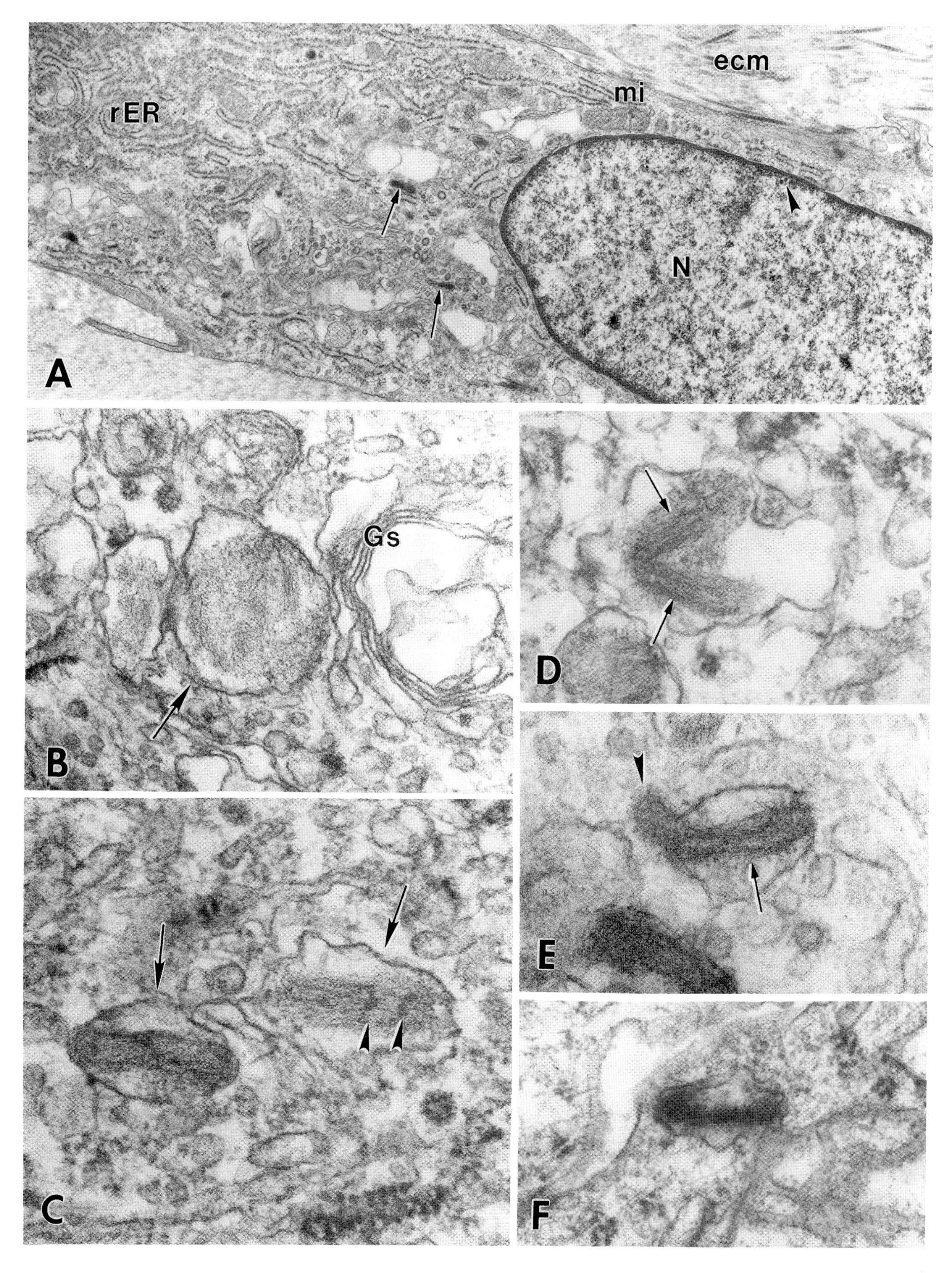
ecm
mi
rER
N
A
Gs
B
C
D
E
F

tous bundle reorganizes into longitudinal rodlets (Plate 30A); these are solid and thus can be distinguished from the **microtubular inclusions** of Weibel–Palade bodies. Finally, the interior becomes homogeneously dense (Plate 30B).

Microtubules are believed to guide collagen secretion granules to the cell periphery (Plate 30C), where the limiting membrane fuses with the plasmalemma. The secreted procollagen is then enzymatically converted to collagen which crystallizes into the familiar cross-striated fibrils.

INTRACELLULAR COLLAGEN

While collagen fibrils normally form outside cells, they *can* be seen apparently within cytoplasm, sometimes enclosed by a membrane. In some of these instances, the fibrils are almost certainly *not* within the cytoplasm; they merely appear to be so, and in fact they are in depressions of the cell surface (Plate 30D). Sometimes, a limiting membrane is not obvious. This may be due to the fact that the membrane of the cell surface depression is at an angle to the section and, not being sharply imaged, *appears* to be absent; or the membrane is indeed absent, having broken down.

It can be difficult also to know whether collagen fibrils surrounded by membrane are newly formed and in a shallow bay, or being endocytosed. Some ultrastructural appearances can indicate endocytosis, which is known to take place in a variety of mesenchymal cells: fibroblasts, myofibroblasts, osteoblasts, and chondrocytes (macrophages also ingest collagen). For example, when collagen fibrils are encircled by a membrane which also encloses a dense granular matrix or other structures such as vesicles, this suggests macromolecular breakdown and a lysosomal organelle (Plate 30E and 30F).

Plate 30. **A**: Collagen secretion granule containing dense rodlets (*arrow*) and with coated vesicle budding off membrane (*arrowhead*). Same tissue as in Plate 29E. ×98,000. **B**: Mature collagen secretion granule (*arrow*) with homogeneously dense interior and a hint of polar density. A coated vesicle is nearby (*arrowhead*), which is itself surrounded by dense glycogen granules (beta-type; see Chapter 20). Myofibroblast in foreign body granuloma. ×95,200. [**A** and **B** reproduced from Eyden (1989), with permission of Springer-Verlag.] **C**: A dense collagen secretion granule (*arrow*) near the cell surface. Note microtubules. Stromal myofibroblast in primary infiltrating duct carcinoma. ×20,000. **D**: Numerous channels or vacuoles containing fibrils. Some collagen fibrils in the matrix have enlarged diameters (*arrow*). [Micrograph courtesy of Professor Germana Dini (Florence). Reproduced from Dini et al. (1986), with permission from *Journal of Submicroscopic Cytology and Pathology*.] ×40,000. **E**: Collagen fibrils (*arrow*) and microvesicles (*arrowhead*) in an endocytotic vacuole. Reactive myofibroblast in epithelioid sarcoma of scalp. ×60,500. **F**: Pale-staining collagen fibril (*arrow*) associated with dense material (*) in a secondary lysosome. Note narrow space of uniform width under membrane (*arrowheads*), which is a typical lysosomal feature. Normal endometrial stromal cell. ×70,700. ecm, extracellular matrix; g, glycogen; mt, microtubules.

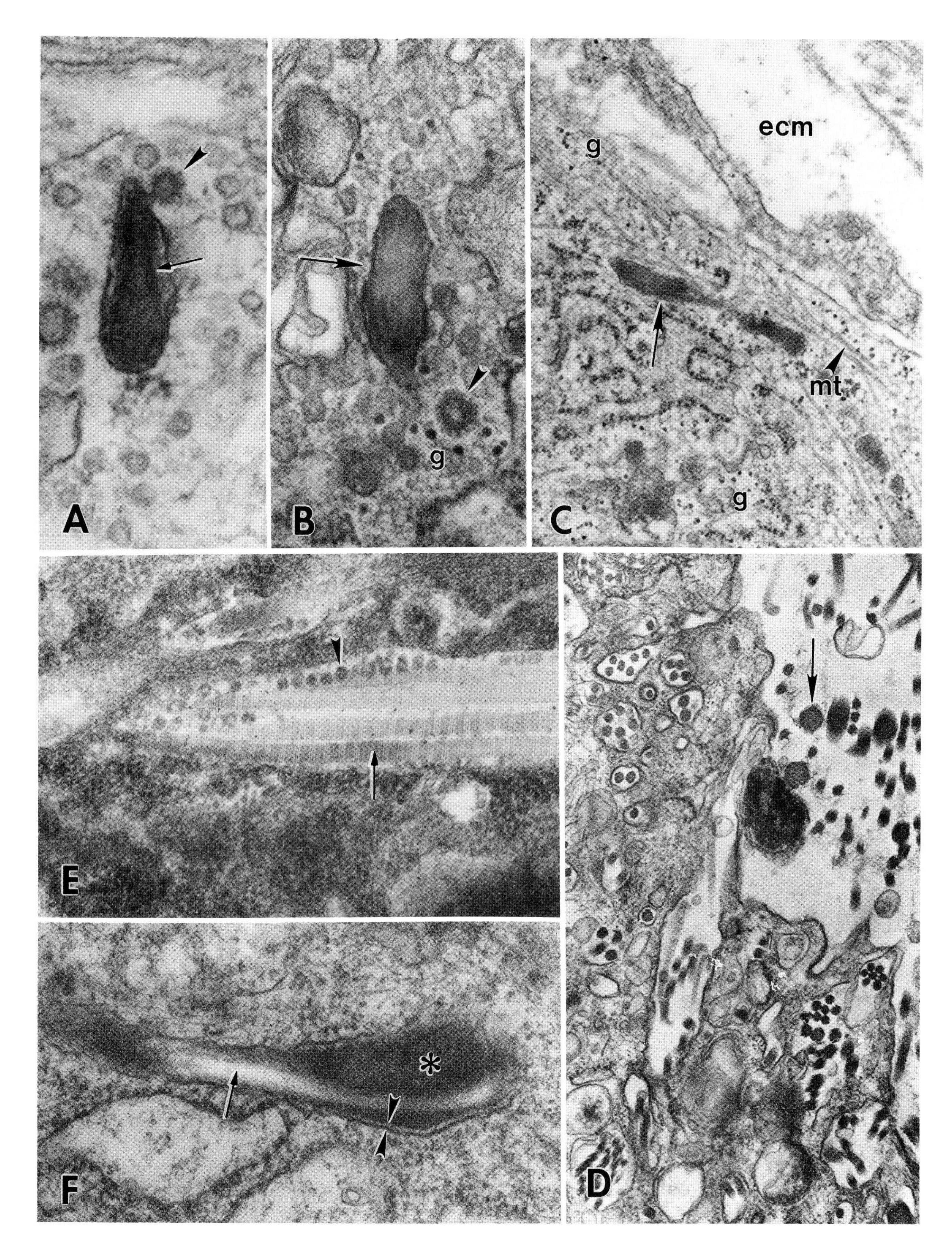
A
B
C
D
E
F
g
ecm
mt
*

13

Endocytosis and Secondary Lysosomes

In contrast to primary lysosomes (which have *not* taken part in digestion), **secondary lysosomes** are organelles in which digestion either is in progress or has taken place. They are identified by enzyme cytochemistry for hydrolases and/or by the ultrastructural evidence of decomposing cell or matrix constituents.

Secondary lysosomes are almost ubiquitous in cells, but they can be numerous in: myeloid, monoblastic, and monocytic leukemias (including granulocytic sarcoma), true histiocytic lymphomas, prostatic epithelial tumors, neuroendocrine and steroidogenic tumors, and Schwann cell neoplasms. They are most abundant, however, in macrophages and granular cell tumors.

Secondary lysosomes are formed when **primary lysosomes** containing hydrolases fuse with membrane-bound bodies containing the material to be digested. This material can be particulate and of external origin (bacteria, cells, matrix fibrils), particulate and of cytoplasmic origin (organelles), or in the form of dispersed molecules ("fluid") from the extracellular milieu. Different terms are used for the organelles which prepare and present the material to primary lysosomes for digestion depending on their origin and nature; these will be described first (Chapters 13–16), followed by the secondary lysosomes themselves (Chapter 17).

ENDOCYTOSIS (PHAGOCYTOSIS AND PINOCYTOSIS): FUNCTION AND TERMINOLOGY

The term **endocytosis** is used for all forms of uptake of exogenous material. **Particulate exogenous material** is taken up by **phagocytosis**, and the membrane-bound

Plate 31. **A**: Phagocytosed cell (*arrows*) in a well-preserved tumor cell. Note that despite decomposition, the nucleus and mitochondria (*arrowhead*) are still identifiable. Hodgkin's disease, axillary node. ×8400. **B**: Autophagic vacuoles. The fact that the enclosed mitochondria (*) seem relatively unaffected structurally but that the surrounding space has some vesicular structures (*arrowheads*) suggests that these might be very early secondary lysosomes. Granular cell tumor, palm. ×49,800. **C** and **D**: Coated invagination (**C**) and coated vesicle (**D**). Benign schwannoma, foot. Both ×95,200. hc, heterochromatin; mi, mitochondria; N, nucleus; Nu, nucleolus; pm, plasma membranes.

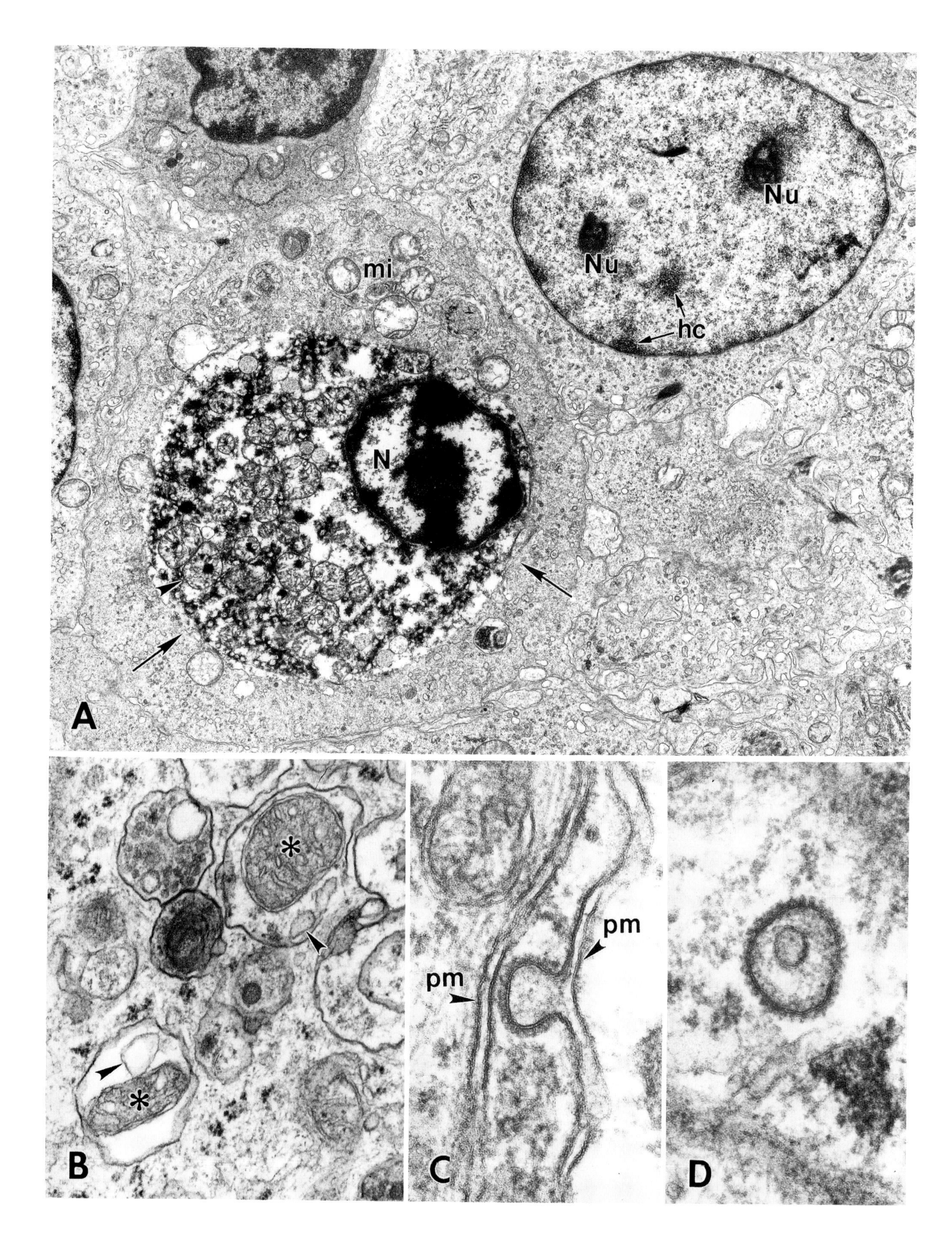
Nu
Nu
mi
hc
hc
N
A
*
*
pm
pm
B
C
D

body containing the phagocytosed item is a **phagosome** or, to emphasize its exogenous origin, a ***hetero-phagosome***. Examples include bacteria ingested by granulocytes and macrophages, and erythrocytes and other cells ingested by macrophages or tumor cells (Plate 31A). When the heterophagosome fuses with a primary lysosome to initiate digestion, it becomes a **secondary lysosome**. More scientifically precise terms exist in the literature—*heterophagolysosome* and *heterolysome*—but secondary lysosome is preferred.

Cytoplasmic organelles are prepared for digestion by being first encapsulated in an **autophagic vacuole** or **autophagosome** (Plate 31B); an older term is *cytosegresome*. When this fuses with a primary lysosome, it produces a secondary lysosome. Again, the terms *autolysosome* (*autolysome*) can be found in the literature, but secondary lysosome is preferred.

In purely ultrastructural terms, it may sometimes be difficult to know whether a given organelle is merely a phagosome or whether it is a secondary lysosome—that is, to know whether digestion has been initiated. Without the cytochemical or immunocytochemical demonstration of a hydrolase, one infers a secondary lysosome from structural evidence of decomposition (Plate 31A).

Cells can internalize other cells without digesting them; this is **emperipolesis** (Takeya and Takahashi, 1988).

PINOCYTOSIS

The internalization of dispersed molecules occurs by **pinocytosis**; here, two structures are identifiable. One involves areas of membrane coated with the protein **clathrin**, which invaginates to form a **coated pit** or **caveola** (Plate 31C). Detachment produces a **coated vesicle** (Plate 31D); these are ~80–250 nm across. Obsolete terms encountered in the literature include *vesicle-in-a-basket*, *acanthosome*, and *spiny vesicle*. Coated vesicle formation is also known as **receptor-mediated endocytosis**, because the membrane involved possesses receptors for certain kinds of ligand. Therefore, coated vesicles have also been called *receptosomes*.

The clathrin coat is frequently seen in cells where there is bulk flow of membrane, as in endocytosis or the pinching-off of membrane in early post-Golgi secretory vacuoles; here, the process enables secretion to be concentrated and the vacuole to become smaller (see Plates 9A, 27D, and 29E).

PINOCYTOSIS INVOLVING SMOOTH MEMBRANE

A second mode of pinocytosis uses *smooth* membrane. The cell surface invaginates to form **smooth-membraned pits** or **caveolae**; then, complete **pinocytotic vesicles (pinosomes)** detach and migrate into the cytoplasm. Because these vesicles are small (60–80 nm), the process is often referred to as *micropinocytosis*, but the term *pinocytosis* is just as suitable.

While coated pits and coated vesicles derived from cell surface membrane are often found in relatively small numbers (they can be numerous in the Golgi region, however), smooth-membraned pinocytotic caveolae and pinocytotic vesicles are abundant in endothelial cells (Plate 32A). Here, once internalized they tend to aggregate together to produce floret configurations (Plate 32C). It is important to observe these vesicles in relation to a vertically sectioned surface membrane before judging them to be truly intracellular (Plate 32C). Plasmalemmal caveolae are often referred to as pinocytotic *vesicles*. In cells showing smooth-muscle differentiation in particular, one often sees rows of such caveolae (Plate 32B). Strictly speaking, these are not vesicles because they are still attached to the surface membrane; there is also no evidence that they are in fact pinocytotic. Therefore, in this situation, **plasmalemmal caveola** is preferable to *pinocytotic vesicle*.

Pinocytotic vesicles can survive paraffin embedding.

Plate 32. **A**: Pinocytotic vesicles and pinocytotic caveolae in a vessel from aggressive angiomyxoma of the broad ligament (see Plate 27A for a low-power perspective of a vessel). Some pinocytotic vesicles are fusing together (*small arrow*), while the luminal surface is invaginating to provide more surface for pinocytotic activity (large arrow). Asterisks (*) indicate interendothelial cell junctions (tight junctions) which in these instances are poorly defined partly because of sectioning geometry. Note also the finely textured extracellular matrix which contrasts with the structureless lumen. ×60,000. **B**: Rows of plasmalemmal caveolae (*arrows*) in epididymal smooth-muscle cells. Note densely staining collagen fibrils in matrix, and also note attachment plaques (*arrowheads*). ×45,400. **C**: A floret arrangement of pinocytotic vesicles (*arrow*) in a tumor cell from myofibrosarcoma, axilla. Note pale-staining collagen in matrix. ×39,300. co, collagen fibrils; ecm, extracellular matrix; my, smooth-muscle myofilaments; pv, pinocytotic vesicles.

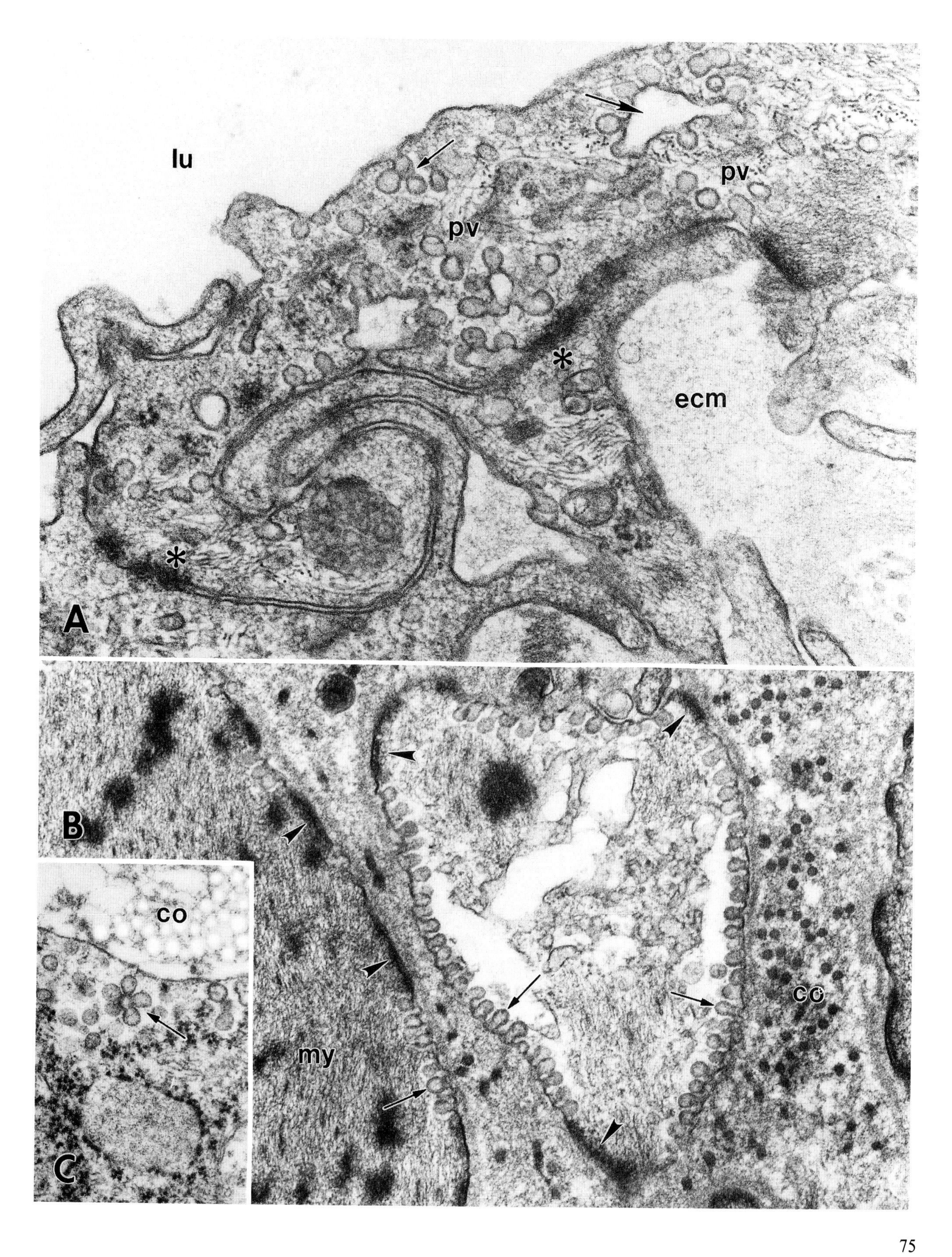
lu
pv
pv
ecm
A
B
co
co
my
C

14

Langerhans Cell (Birbeck) Granules

The **Birbeck granule** is highly specific for the Langerhans cell (Plate 33A)—hence the alternative and preferable name of **Langerhans cell granule**. It is believed to be endocytotic, with the role of taking up foreign antigen for processing within the lysosomal system before secreting it for the activation of other elements of the immune response.

ULTRASTRUCTURE OF LANGERHANS CELL GRANULES

Langerhans cell granules are membrane-limited and appear in section as rather short and narrow rod-shaped "tubules" with a periodic line of dot-like densities running down the central enclosed space (Plate 33B); they are *not* tubules, however, but discoid (Plate 33E). Typically, they are ~40 nm across in profile and ~200–400 nm long. They may be found solitarily or in small groups. So-called *tennis racquet* forms have a clear expansion at one pole (Plate 33C). They may show a membranous constriction (Plate 33B and 33E), areas of clathrin coat, and sometimes a square lattice (Plate 33D). Clathrin-coated Langerhans cell granules can be seen in continuity with the plasmalemma (Plate 33A, inset), supporting the idea that they are endocytotic.

Less often, Langerhans cell granules have curved or annular profiles (Plate 33E), sometimes near or attached to the plasmalemma. Structures which have a similar appearance but which may not be true Langerhans cell granules have also been described (El-Labban, 1982) (Plate 33F). Langerhans cell granules can survive paraffin embedding and can be found in exceptionally poorly preserved specimens (Dardick et al., 1993).

Plate 33. **A**: Typical distribution of Langerhans cell granules (*arrows*). The circled granule is detailed in the inset. ×8600. **Inset**: detail showing clathrin coat (*arrowhead*). ×98,600. **B**: Langerhans cell granule showing central dotted "line" and constriction (*arrow*). ×143,000. **C**: Tennis racquet form with expanded polar vesicle (*arrow*). ×98,000. **D**: Square lattice of Langerhans cell granule (*). ×83,300. **E**: Langerhans cell granules showing circular (1) and discoid (2) appearances, and one with a constriction (3). ×95,200. **F**: Annular profile mimicking Langerhans cell granule. Note the difference from **E** in that the internal material (*arrowheads*) is thick and amorphous and not dot-like, and that the diameter varies. Reactive macrophage. ×150,000. **A–E** are of Langerhans cell granulomatosis (histiocytosis X); **A**, **B**, and **E** are from a submental node; **D** is from the ischial ramus; ly, secondary lysosome; N, nucleus; Nu, nucleolus; [**C** courtesy of Irving Dardick, M.D. (Toronto). **A**, **B**, and **C** reproduced from Dardick I. *Handbook of Diagnostic Electron Microscopy for Pathologists-in-Training*, New York, Igaku-Shoin, 1996, with permission.]

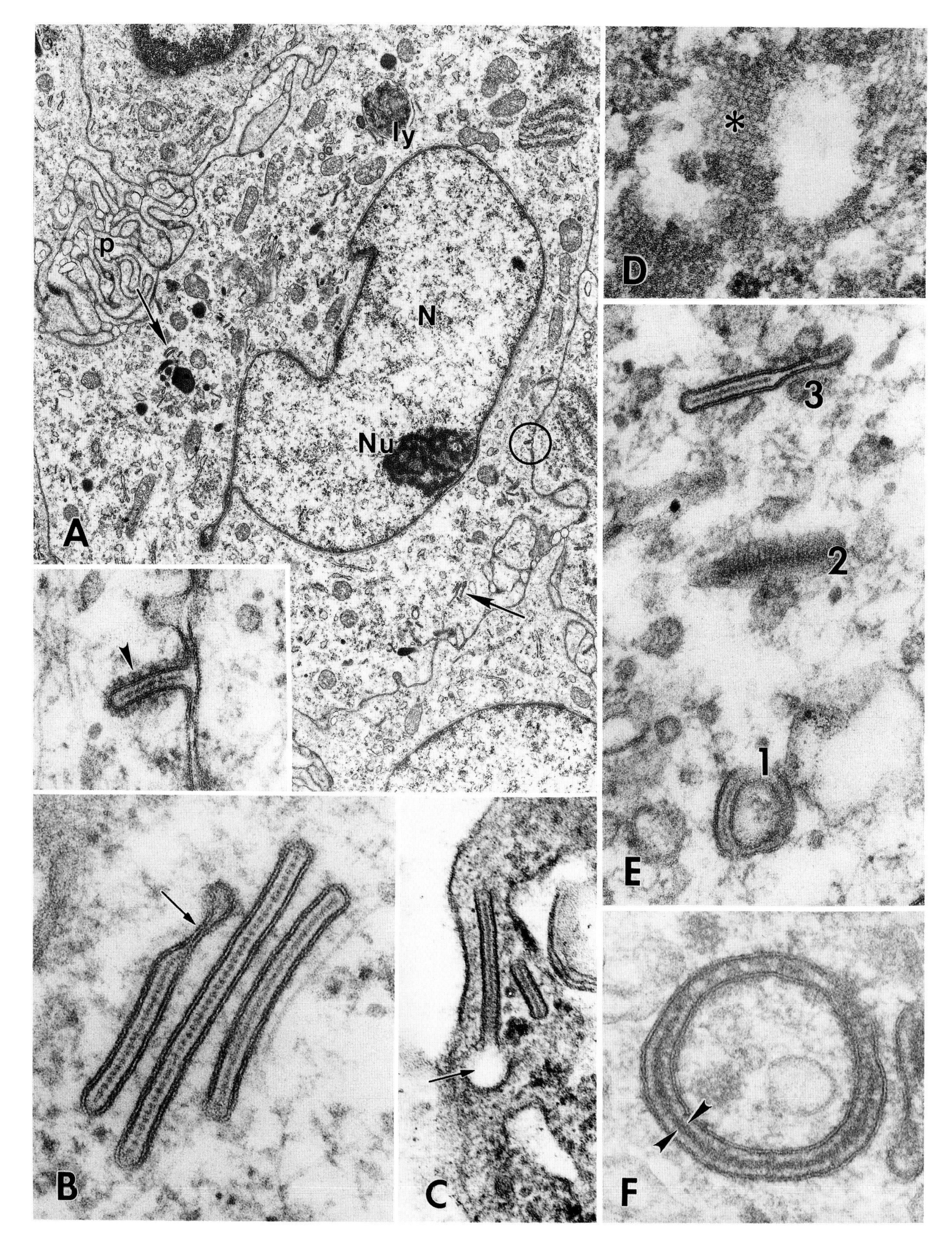
ly
p
N
Nu
A
D
3
2
1
E
B
C
F

15

Multivesicular Bodies

Multivesicular bodies (MVBs) are probably produced as a result of the fusion of pinocytotic vesicles. They are sometimes referred to as *endosomes*.

One or two MVBs are found in most cells, but they predominate in intestinal and other luminal absorptive cells. They also occur in true histiocytic lymphomas and in the vacuolar variant of signet-ring cell lymphomas.

THE ULTRASTRUCTURE OF MVBS

MVBs are single-membraned bodies containing small, rounded membranous vesicles (also called **microvesicles**) (Plate 34A). The vesicles are ~60–80 nm in diameter, and the MVBs are ~200–800 nm across. Clathrin-like material may be seen on the MVB membrane in an area which also seems to be stiffened (Plate 34B). MVBs are often surrounded by small clear vesicles having the same appearance and size as the vesicles inside the MVB (Plate 34A and 34B). They may represent either clear primary lysosomes or acidifying vesicles, since the MVB is known to contain acid phosphatase.

Some MVBs just contain vesicles, whereas others have vesicles mixed with disorganized material, suggesting digestion (Le Tourneau et al., 1990). Digested solutes diffuse out of the MVB into the cytosol. Normally, membrane would be pinched off the MVB for recycling, but sometimes in tumors one sees **reticulate MVBs** with an exceptionally large surface area, suggesting a failure in membrane recycling (Plate 34C and 34D).

The inclusion responsible for the signet-ring appearance in some lymphomas resembles a giant vacuole-like MVB or **multivesicular vacuole** (Plate 34E). In some of these lymphomas, the inclusion lacks vesicles and is therefore more appropriately referred to as a **vacuole** (Plate 35).

Plate 34. **A:** Typical size and shape of MVBs. One MVB (*arrow*) is surrounded by vesicles (*arrowheads*). T-cell lymphoma, cervical node. ×30,000. **B:** Pale MVB with clathrin-like coat (*arrow*) along a straight portion of membrane (*). Note surrounding vesicles (*arrowheads*). Smooth-muscle cell from uterine leiomyoma. ×70,700 **C:** Two dark MVBs (*arrows*) in different stages of development. The one indicated by *—detailed in **D**—may be showing a membrane-recycling abnormality. **C**, ×14,000; **D**, ×78,000. [Both figures reproduced from Eyden et al. (1990), with permission from Springer-Verlag.] **E:** Multivesicular vacuoles containing peripheral microvesicles and central empty space. ×5300. **C–E** are cutaneous T-cell lymphoma of vacuolar signet-ring cell type (Cross et al., 1989; Eyden et al., 1990). ecm, extracellular matrix; hc, heterochromatin; N, nucleus; np, nuclear pore; Nu, nucleolus; v, vacuole.

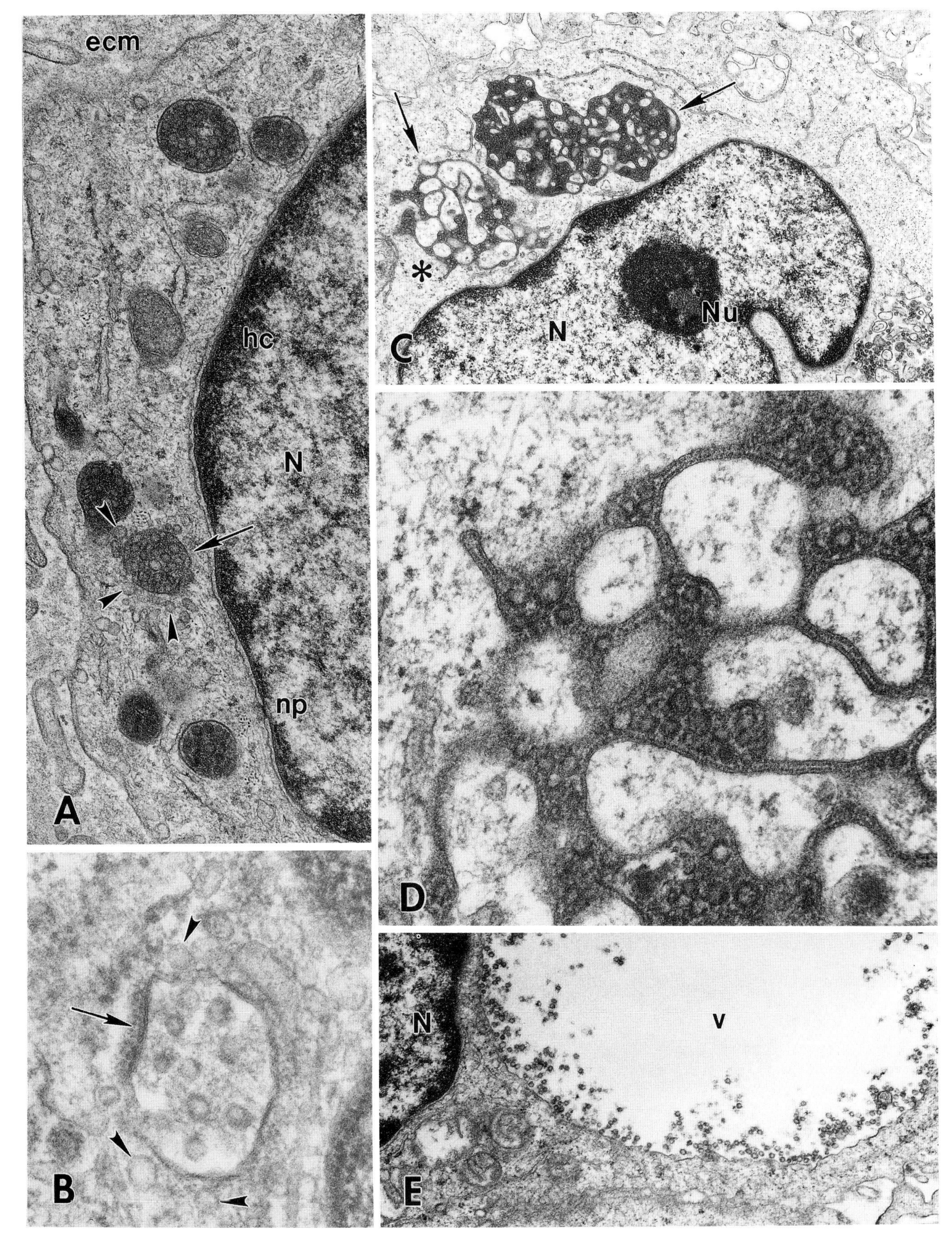
ecm
hc
N
np
A
B
*
Nu
N
C
D
N
V
E

16

Vacuoles

Vacuoles are rounded cytoplasmic inclusions defined by a single membrane and a clear structureless content. Small vacuoles (100–200 nm) may be referred to as vesicles. Vacuolar contents are generally of undefined composition.

In spite of their largely unknown nature, vacuoles have some diagnostic value. Small to medium-sized vacuoles may be numerous in malignant melanoma (where they seem to be abortive melanosomes), some clear-cell tumors, and certain carcinomas [e.g., chromophobe cell variant of renal cell carcinoma (these may be endocytotic and analogous to the vesicles of the collecting duct epithelium of the normal kidney)]; in mesenchymal tumors such as atypical fibroxanthoma (Plate 35A) and in multinucleated giant cells (Plate 17C and 17D) (where they may be abortive lysosomes); and in nerve sheath, myxoid, and epithelioid sarcomas. Larger vacuoles (micrometer range and above) can be found in vasoformative tumors, mesotheliomas, and some signet-ring cell lymphomas (Plate 35B) and stromal tumors of the ovary.

The following points are worth making with regard to vacuoles and vacuole-like structures: (1) Some vacuoles or vacuole-like cisternae have apparent continuity with cell surfaces and sometimes have a content indistinguishable from the extracellular matrix (Plate 35C). This suggests that these vacuolar profiles may be deep labyrinthine intrusions of matrix into the cell. (2) Cells appearing vacuolated by light microscopy may on ultrastructural examination contain lipid droplets or true membrane-lined vacuoles. Light microscopically observed vacuolation may also be due to artifactually distended organelles such as mitochondria and rough endoplasmic reticulum (rER) (Ishimaru et al., 1988). (3) Vacuoles in angiosarcoma may represent primitive lumina. Some structures interpreted as vacuoles, however, are in fact distended rER because of the presence of ribosomes on the membrane (Yang et al., 1981).

Many vacuoles survive paraffin embedding, but the vesicles of chromophobe renal cell carcinoma do not (Bonsib et al., 1993).

Plate 35. **A**: Numerous clear vesicles/vacuoles 300–400 nm across (for details of these vesicles, see Plate 12B). Atypical fibroxanthoma, skin of ear. ×5500. **B**: Single giant vacuoles, 7 μm across. Signet-ring cell lymphoma, femoral node. ×5900. **C**: Vacuole-like profiles (*) which could represent either endocytotic structures or intrusions of matrix into the cytoplasm. Ganglioneuroma, cervical node. ×13,300. ecm, extracellular matrix; N, nucleus; Nu, nucleolus; rER, rough endoplasmic reticulum; V, vacuole.

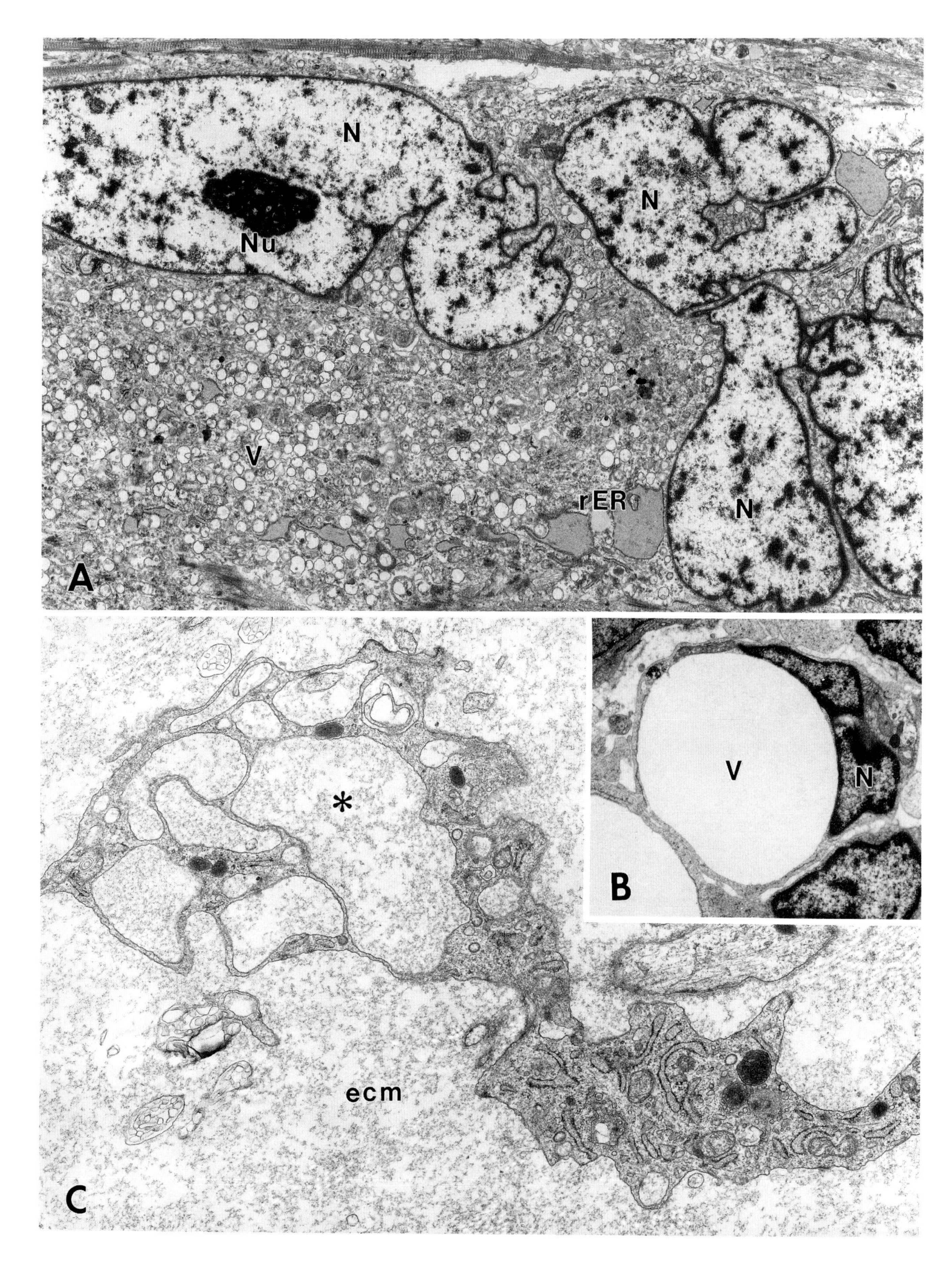
N
N
Nu
V
rER
N
A
V
N
B
*
ecm
C

17

Secondary Lysosomes

Typical **secondary lysosomes** have a single membrane and are characterized by a disorganized content consisting of electron-dense amorphous material, lamellae or membranes, microvesicles, or granules (Plate 36A and 36B). Sometimes, cellular constituents such as neuroendocrine granules, melanosomes, and glycogen may be recognizable. Secondary lysosomes can be up to several micrometers across.

SECONDARY LYSOSOMES ARISING FROM ERYTHROPHAGOCYTOSIS

Erythrophagocytosis is encountered in a wide range of tumors (Plate 36C) and particularly in Kaposi's sarcoma, as well as in macrophages in situations of hemorrhage such as hemophilic arthropathy and villonodular synovitis. The ingested erythrocytes are recognized by their distinctive density and extremely fine and uniform texture (Plate 36C). In some instances, the digestion of erythrocytes seems to be followed by a fragmentation into material localized within small vacuoles (Plate 36C). In other instances, one can see a margination of dense material at the periphery of the ingested erythrocyte (Plate 37A and 37B). As the digestion of erythrocytes proceeds, **siderosomes** are formed (Plate 37C) which are the ultrastructural counterpart of **hemosiderin pigment**. Siderosomes are single-membrane-bound lysosomes containing abundant small iron-containing particles; the latter may be confined to siderosomes, or they may also be found throughout the cytoplasmic matrix (Plate 37C).

LIPID ACCUMULATION IN SECONDARY LYSOSOMES

As digestion proceeds in secondary lysosomes, indigestible residues remain, some of which are lipids. This presents as rounded droplets varying greatly in number,

Plate 36. **A**: Large numbers of predominantly dense secondary lysosomes in the cytoplasm. Note collagenous septum, and also note interstitial cell containing angulate bodies (see Plate 39A and 39B). Granular cell tumor, breast. ×5100. **B**: Detail of secondary lysosome. Same tumor as in **A**. ×22,000. Reproduced from Dardick I. *Handbook of Diagnostic Electron Microscopy for Pathologists-in-Training*, New York, Igaku-Shoin, 1996, with permission. **C**: Large and small fragments of ingested erythrocyte material (*) in the cytoplasm of a large cell. Note the tiny nuclear profile and the adjacent fibroblast. Malignant histiocytosis, axillary and cervical skin. ×10,200. co, collagen; f, fibroblast; ic, interstitial cell, N, nucleus; Nu, nucleolus.

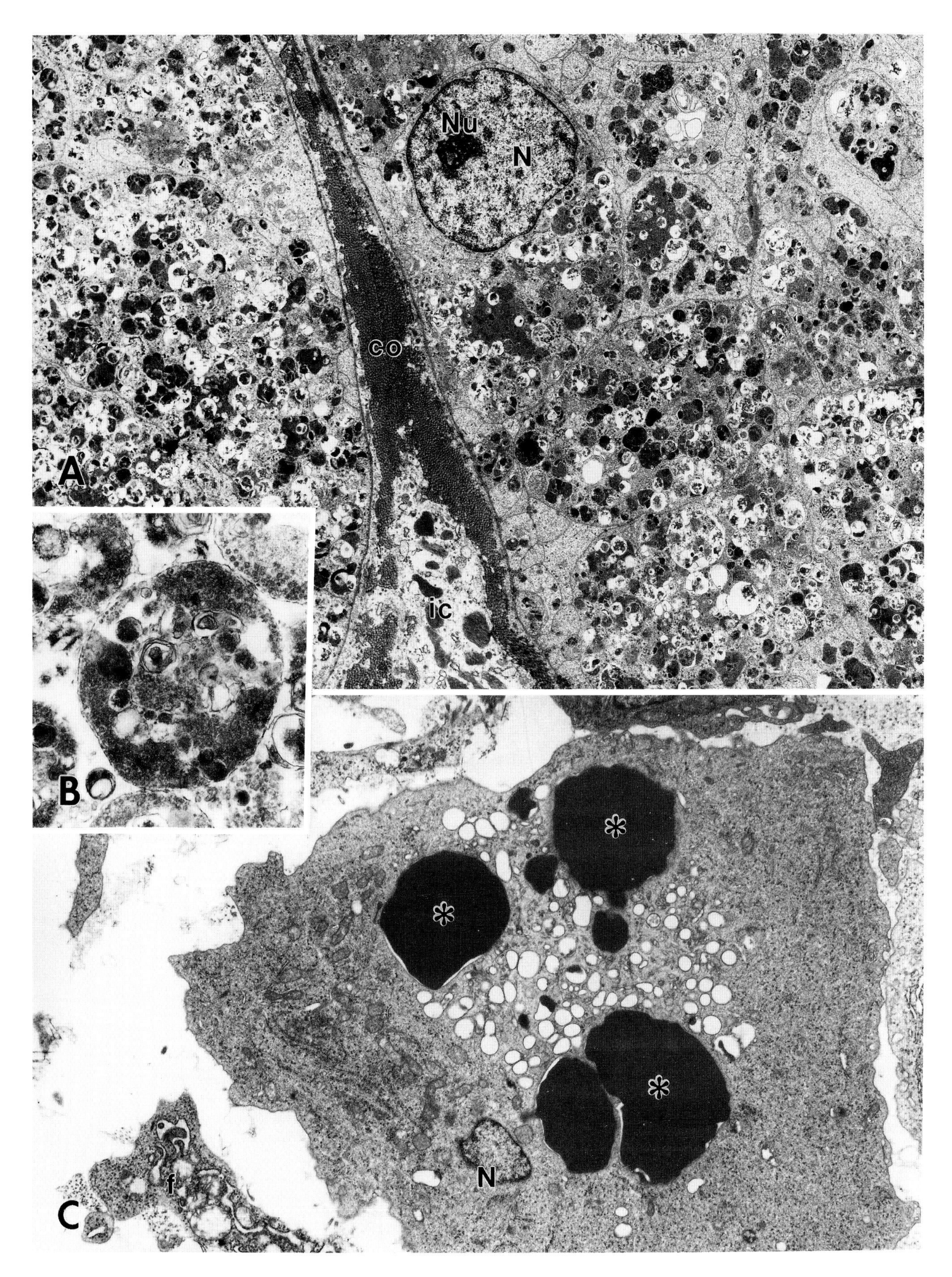
Nu
N
co
A
ic
B
C
f
N

size, and electron density. At one extreme, one may see only a solitary lipid droplet (see Plates 26C, 37C, and 39B); at the other, one may see a large number of different sizes and densities (Plate 38A). Secondary lysosomes with substantial lipid droplet accumulation but also with a significant proportion of dense granular background matrix are referred to as **lipofuscin granules** (Plate 38B) (see also Plate 11B). Lipofuscin granules are commonly encountered in neuroendocrine and steroidogenic tumors, including the black adrenal adenoma.

Bodies which are demonstrated to be secondary lysosomes by their content of hydrolases and which are almost but not quite full of lipid are referred to as **lipid-rich residual bodies** or **lipolysosomes**. They are nonspecifically distributed among epithelial and mesenchymal cells and tumors: renal oncocytoma, normal myometrium and uterine leiomyoma, and thyroid and sweat gland epithelium (Plate 38C). Although their great lipid content disqualifies them as typical lipofuscin, they can be regarded as a lipid-rich variant.

DISTINCTIVE LYSOSOMAL VARIANTS

A number of organelles are recognizable as lysosomes (*e.g.*, by hydrolase cytochemistry), but they lack the typical features of lysosomes as described above. These include: **angulate bodies, parallel-tubular-array lysosomes**, and **Auer rods (Auer bodies)**. The presence of lipid droplets or a disorganized content as in angulate bodies and some Auer rods suggests that they are secondary lysosomes.

ULTRASTRUCTURE

Angulate bodies (Plate 39A and 39B) are found in macrophages and stromal cells in granular cell tumors. They have angular profiles and contain fine filaments. These are often in pairs, appear to be laminae, and might be lipid since they have a sharper delineation than membrane (Mooi et al., 1983). These organelles are distinct from Gaucher's bodies, where internal circular profiles have been observed (~30–60 nm in diameter) and where, therefore, tubules are present (Lee and Ellis, 1971).

Lysosomes containing parallel tubular arrays (PTA-lysosomes) are found in certain classes of lymphocyte. They have a single limiting membrane enclosing tubules ~30–50 nm diameter (Plate 39C and 39D). These lie in a clear space in which acid phosphatase has been demonstrated.

Auer rods (Auer bodies) are found in acute myeloid leukemia. They are elongated bodies, several micrometers long and therefore readily observed by light microscopy. They contain filamentous or tubular elements, or cross-striated fibrils (McDuffie, 1967; Mintz et al., 1973; Stavem et al., 1981).

Further lysosome variants include: **juxtaglomerular cell tumor** and **alveolar soft-part sarcoma crystals** (Chapter 29); the **Michaelis–Gutmann body**, which is associated with malakoplakia and which has a concentric substructure partly involving calcification (August et al., 1994); **Paneth cell granules**, which lack striking ultrastructural features but are distinctive in being found in Paneth cells of the intestinal crypts (Dvorak and Dickersin, 1980); and the **hyaline bodies** in embryonal sarcomas of liver (Chapter 9).

Plate 37. **A** and **B**: Decomposing ingested erythrocytes (*) with marginal dense material (*arrowheads*) and associated dense siderosomes (*arrows*). Kaposi's sarcoma, inguinal node. Both ×21,100. **C**: Iron-containing particles concentrated within siderosomes (*) but also lying in the cytoplasmic matrix (some examples are circled). Note the uniform submembranous halo (*arrowheads*) (typical of lysosomes) and the lipid (typical of secondary lysosomes). Macrophage in granulation tissue. ×95,200. Go, Golgi apparatus; li, lipid; N, nucleus; v, vacuole.

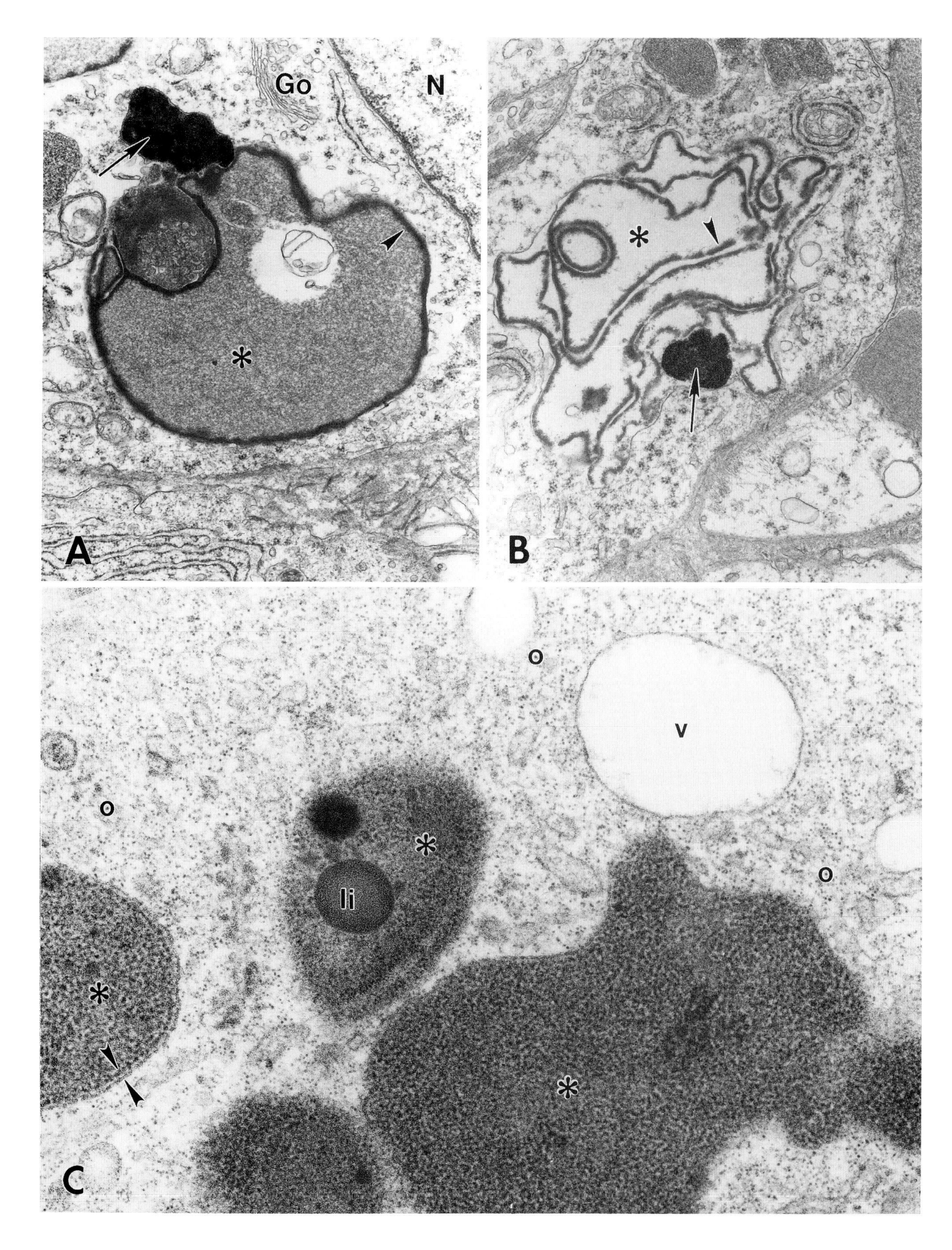
Go
N
*
A
*
B
o
v
o
*
li
o
*
*
C

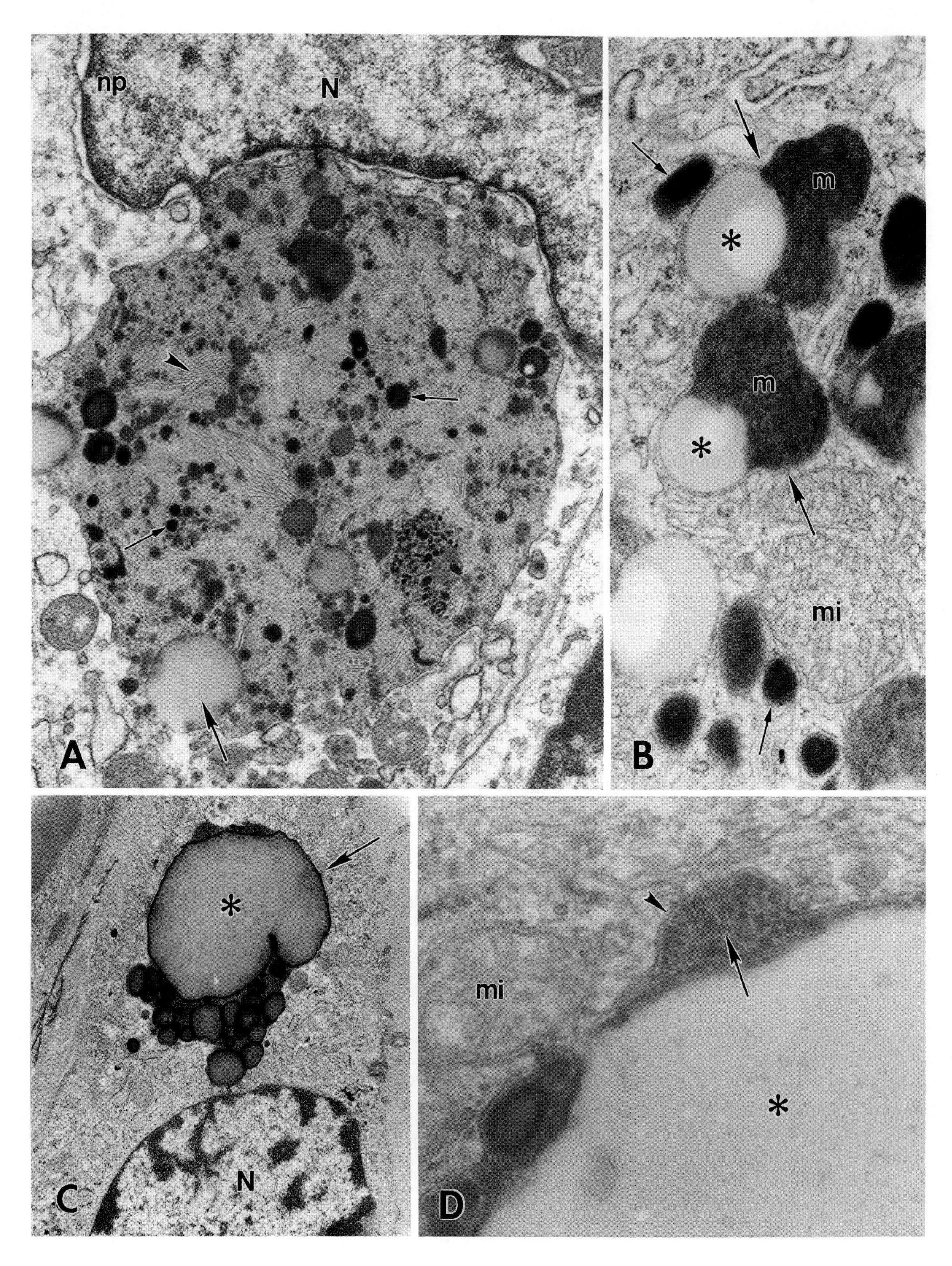
np
N
A
m
m
mi
B
C
N
mi
D

Plate 38. **A**: Single large lysosome indenting the nucleus in a sinus histiocyte containing dense lipid droplets of varying size (*small arrows*), a smaller number of larger, paler lipid droplets (*large arrow*), and some foci of filaments or lamellae (*arrowhead*). × 16,900. **B**: Lipofuscin granules (*large arrows*), each containing a single large lipid droplet (*) but also a significant amount of dense, coarsely granular matrix. Pale-staining mitochondria with tubulovesicular cristae, as well as small dense bodies (possibly primary lysosomes) (*small arrows*), are also present. Normal human adrenal cortex. [From a block supplied by Professor G. G. Nussdorfer (Padua). × 38,900.] **C**: Lipid-rich residual body (*arrow*) at the nuclear pole of a cell. The main and the accessory lipid droplets are enclosed by a membrane not resolved at this magnification (see **D**). Normal human thyroid. × 12,000. **D**: Detail of lipid-rich residual body showing the main mass of lipid (*) and peripheral coarse granular lysosomal matrix (*arrow*) within limiting membrane (*arrowhead*). Normal myometrium. × 90,700. [**C** courtesy of Dr. K. Yamazaki (Tokyo). **C** and **D** reproduced from Yamazaki and Eyden (1994) and Eyden et al. (1991), with permission from *Journal of Submicroscopic Cytology and Pathology*.] m, lipofuscin matrix; mi, mitochondrion; N, nucleus; np, nuclear pore.

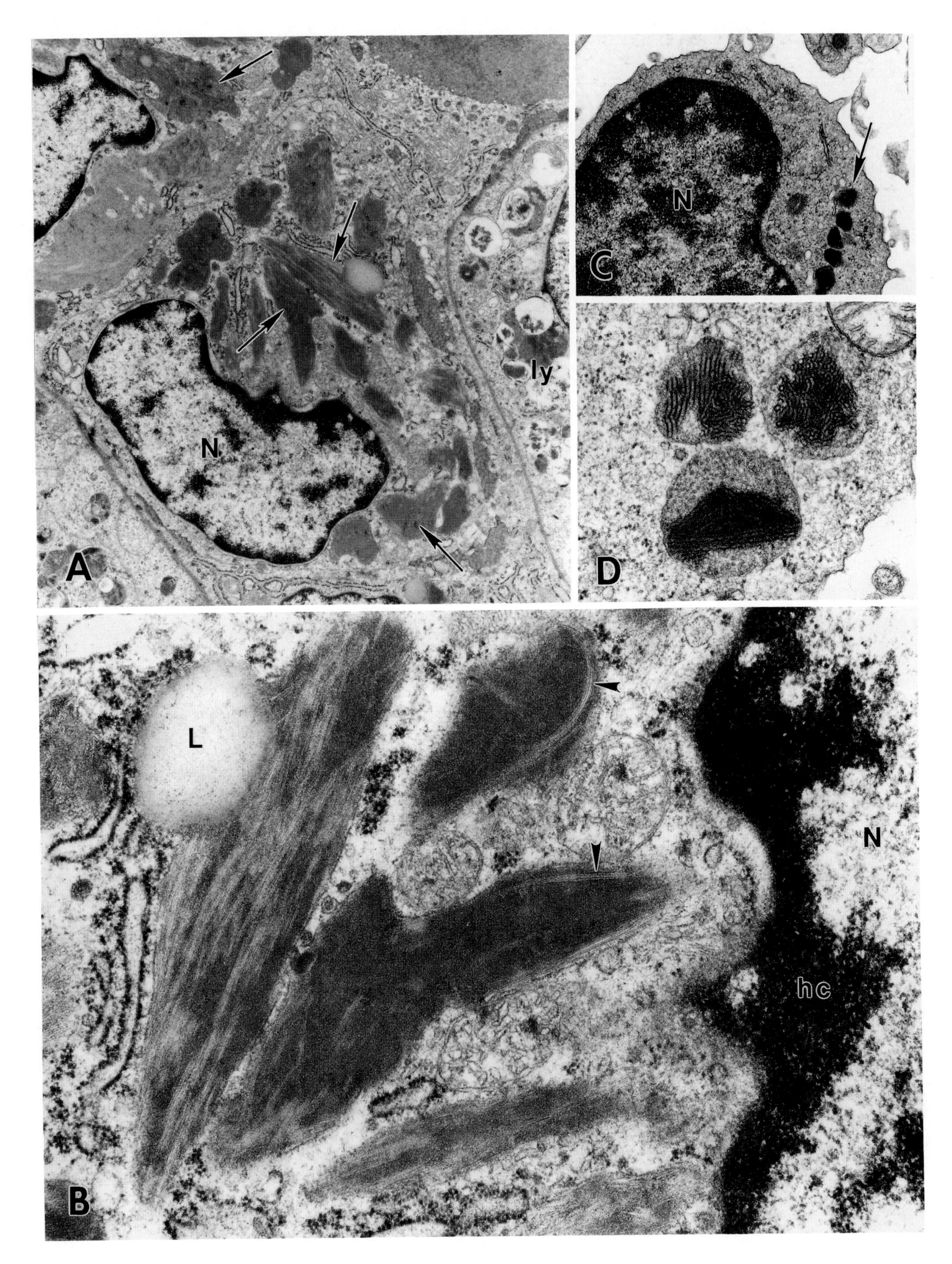
A
N
ly
C
N
D
B
L
N
hc

Plate 39. **A**: Angulate lysosomes (*arrows*) in stromal cells in a granular cell tumor of breast. × 5500. **B**: Detail from **A**. Lipid and laminae (*arrowheads*) are observable. × 34,000 (see Plate 36A for a low-power view of a stromal cell in granular cell tumor). **C** and **D**: PTA-lysosomes (*arrow* in **C**) in a large granular peripheral blood lymphocyte. **C**, × 15,500; **D**, × 44,000. [**C** and **D**: Micrographs courtesy of Ted Wills, M.D. (Campderdown, Australia). Reproduced from Wills et al. (1994), with permission from Taylor and Francis.] hc, heterochromatin; L, lipid; N, nucleus.

18

Mitochondria

Mitochondria, lipid, glycogen, and peroxisomes constitute a functionally related—and to a certain extent physically related—group of cell structures providing energy within cells. Mitochondria are sites of adenosine triphosphate synthesis, while mitochondria and peroxisomes carry out oxidative metabolism. Peroxisomes also possess enzymes for producing and breaking down hydrogen peroxide. Lipid and glycogen are sources of oxidative substrates.

Mitochondria are almost ubiquitous in cells. They are numerous in: steroidogenic cells, some myogenic neoplasms, some renal cell carcinomas; tumors showing *oxyphil* or *oncocytic* differentiation (Plate 40A), including Warthin's tumor; hibernomas; alveolar soft-part sarcomas; hepatocytes and their tumors. A few cells from almost all the major groups of tumors, however, may contain fairly large numbers of mitochondria.

ULTRASTRUCTURE AND IDENTIFICATION OF MITOCHONDRIA

Mitochondria are rounded or ovoid bodies, often ~1 μm across (Plate 40A–C). They may also be rod-shaped (Plate 40A), elongated (Plate 41A), branched or annular. They have the distinctive feature of two limiting membranes (Plates 40C, 41B, and 41C). The outer membrane is smoothly contoured and separated by a narrow clear space from the inner membrane; this forms projections called **cristae**, which are in contact with a finely textured dense matrix (Plate 41).

The matrix may contain a variety of structures. Dense **intramitochondrial granules** include small granules which are nonspecifically distributed and contain lipids or cations (Plate 40B). Large granules, ~300–500 nm across, are of unknown composition and occur in renal cell carcinomas, hepatoblastomas, and adrenocortical carcinomas (Plate 40C). Less sharply defined granules appear as artifacts in poorly preserved mitochondria (Plate 31A). Crystals, glycogen, and myelin figures (Chapter 19) can also occur in the mitochondrial matrix.

The only other cytoplasmic organelles with two limiting membranes include some autophagic vacuoles; therefore, mitochondria are not usually difficult to identify. Although they can show great variations in size and ultrastructural features, if they are reasonably well preserved, the two limiting membranes will provide an unambiguous interpretation. On occasion, however,

Plate 40. **A**: Mitochondria in differing numbers in two cells. (*Above*) Moderate numbers of mitochondria (*arrows*). (*Below*) The numerous and closely packed mitochondria (*) typical of an oncocyte. Mixed glycogen-rich/mitochondria-rich oncocytoma, parotid ×6800. [Micrograph courtesy of Irving Dardick, M.D. (Toronto). Reproduced from Davey et al. (1994), with permission from Mosby-Yearbook.] **B** and **C**: Small (**B**) and large (**C**) intramitochondrial granules (*arrowheads*). Note the two membranes in **C** (*circles*). **B**: Normal mammary epithelium. ×23,300. **C**: Adrenocortical carcinoma. ×39,600. ce, centriole; g, glycogen; lu, lumen; N, nucleus; Nu, nucleolus.

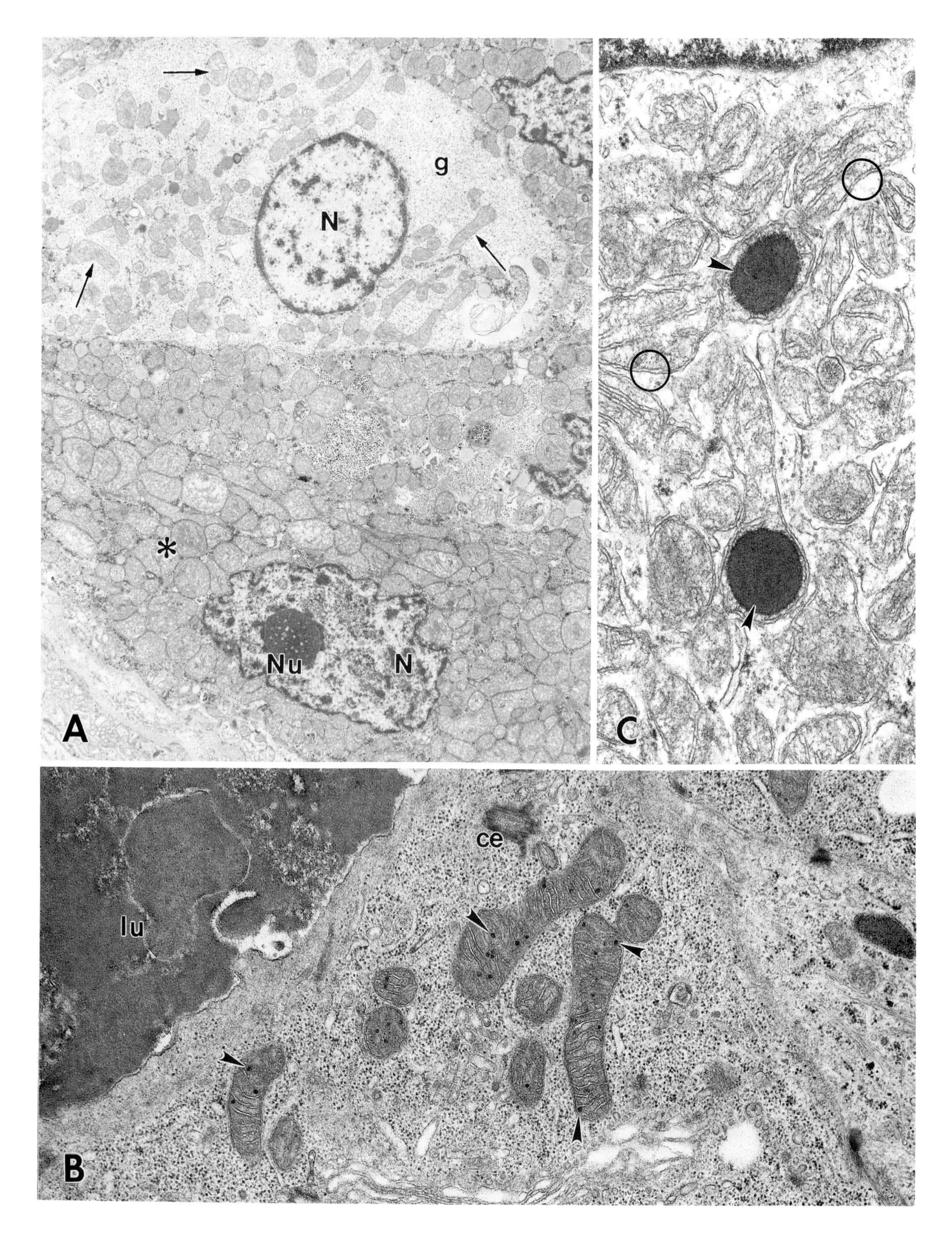
g
N
*
Nu
N
A
C
ce
lu
B

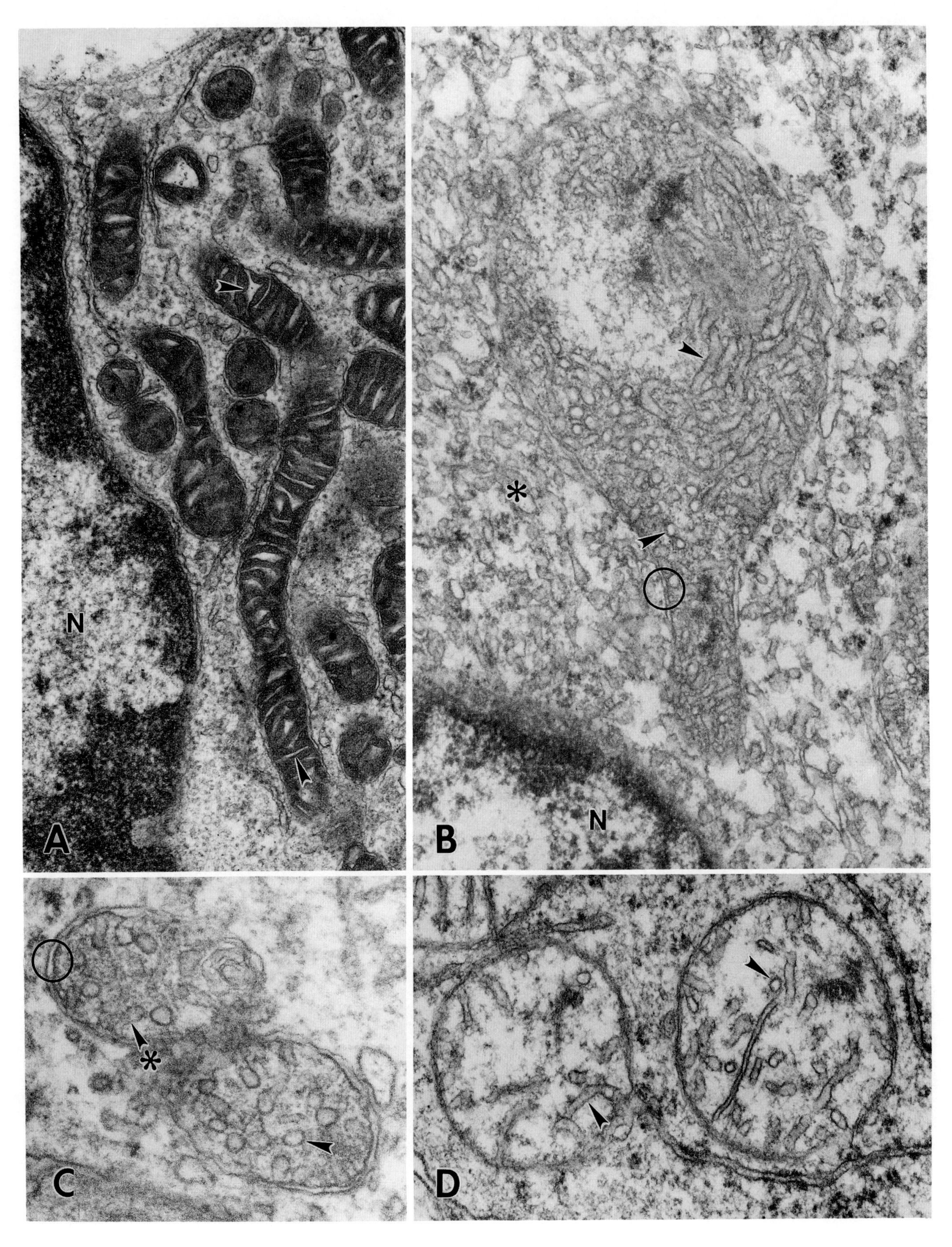
N
A
*
N
B
*
C
D

problems of identification arise. When mitochondria are poorly preserved and the matrix is washed out, they may resemble vacuoles; when tangentially sectioned, the two limiting membranes may not be sharply defined (see Plate 17D); small rounded mitochondria can lack cristae, and if they also possess a very dense matrix, they may resemble neuroendocrine granules (Variend et al., 1991). By contrast, they can also appear pale-staining and resemble mucigen granules (Plate 48A). Mitochondria show variable resistance to paraffin embedding.

Mitochondria are diagnostically important on two main counts. (1) Very large numbers indicate *oncocytic differentiation* (Plate 40A). Here, cristae may form distinctive sheaf-like arrangements (Dardick et al., 1988). (2) Tubular, vesicular, or tubulovesicular cristae, in combination with lipid and smooth endoplasmic reticulum, indicate the steroidogenic phenotype (Plate 41B and 41C). This is typically exhibited by liver, adrenocortical, Leydig, and ovarian cells and their tumors.

Tubular or tubulovesicular cristae also occur in poorly preserved mitochondria (indicated by washed-out matrix) (Plate 41D). Here, it is prudent not to overinterpret them as indicating the steroidogenic phenotype without lipid and smooth endoplasmic reticulum. The interpretation of tubular cristae should also be based, at least in part, on seeing transverse circular profiles and not just elongated finger-like profiles; the latter can equally indicate plate-like cristae (Kay and Armstrong, 1980). Some normal mesenchymal cells and sarcomas also possess tubular cristae in well-preserved mitochondria (Schochet and Lampert, 1978).

Plate 41. **A**: Mitochondria with plate-like cristae (*arrowheads*) and dense matrix. Mantle cell lymphoma, femoral node. ×45,000. **B**: Tubular cristae (*arrowheads*). Steroid cell tumor of ovary. Note vesicles and tubules of smooth endoplasmic reticulum (*). ×39,600. **C**: Cristae in normal adrenal cortex (*arrowheads*). Although mostly vesicular, some have a very short neck (*asterisked arrowhead*) and, thus, are *tubulo*vesicular. Although these mitochondria are suboptimally preserved, they are in an unambiguous steroidogenic cell and therefore the vesicular nature of the cristae is not considered artifactual. ×60,800. In **B** and **C**, the circles indicate the two mitochondrial membranes. **D**: Artifactually tubular cristae (*arrowheads*) in poorly preserved mitochondria (note washed-out matrix). Cutaneous T-cell signet-ring cell lymphoma (Cross et al., 1989). ×55,000. hc, heterochromatin; N, nucleus.

19

Lipid

Triglyceride lipid is found as discrete droplets lying in the cytoplasmic matrix between organelles. It is also often found embedded in or surrounded by glycogen (Chapter 20) and under abnormal conditions in the nucleus.

In small quantities, lipid droplets are almost ubiquitous, but they are abundant in the following: steroidogenic cells and their tumors (e.g., adrenocortical adenomas/carcinomas, steroid cell tumors); lipomas and liposarcomas; sebaceous carcinomas; and renal cell carcinomas.

ULTRASTRUCTURE OF LIPID

Typically, **triglyceride lipid** droplets are rounded. The lipid they contain can be either **amorphous** or **lamellar**. Amorphous lipid has an extremely uniform and very fine texture which very often looks "gray" in micrographs (Plates 38 and 42). It can, however, show a complete spectrum of density depending on chemical composition, degree of extraction by processing reagents, and extent of lipolysis (Plate 42B). In deparaffinized specimens, lipid droplets are usually electron-lucent.

Cholesterol is always completely electron-lucent. It occurs as rounded droplets or rectilinear crystals (Plate 42D).

Lipid droplets in the cytoplasmic matrix are not membrane-bound, although often one can see a "line" at the interface with the cytoplasmic matrix (Plate 42C), or a somewhat thicker rim of dense material (Plate 43B). At low magnification, this can sometimes mimic a membrane. Because of the absence of a membrane, the term lipid *vacuole* should be avoided, as should *liposome*, since this has a widely recognized significance in experimental chemotherapy. Lipid in macrophages may be either free or within lysosomes. Intralysosomal lipid sometimes almost completely fills out the interior of the lysosome and is distinguished from a free cytoplasmic lipid droplet by a true limiting membrane; the lipid here is presumably phagocytosed.

Plate 42. **A**: Multiple lipid droplets of typical appearance (*small arrows*). The large arrows point to rough endoplasmic reticulum (the denser periphery is due to ribosomes). Undifferentiated small-round-cell tumor, breast. ×5500. **B**: Exceptionally dense lipid droplets (*arrows*). Langerhans cell granulomatosis, submental lymph node. ×5700. **C**: Lipid droplet (*) showing absence of membrane at periphery (*arrows*). Note true membranes nearby (*arrowheads*). Cultured human fibroblast from normal lung. ×95,200. **D**: Cholesterol (*arrow*) in a lysosome. Macrophage from same tumor as **B**. ×60,000. co, collagen; N, nucleus; Nu, nucleolus; p, processes; pm, plasma membrane.

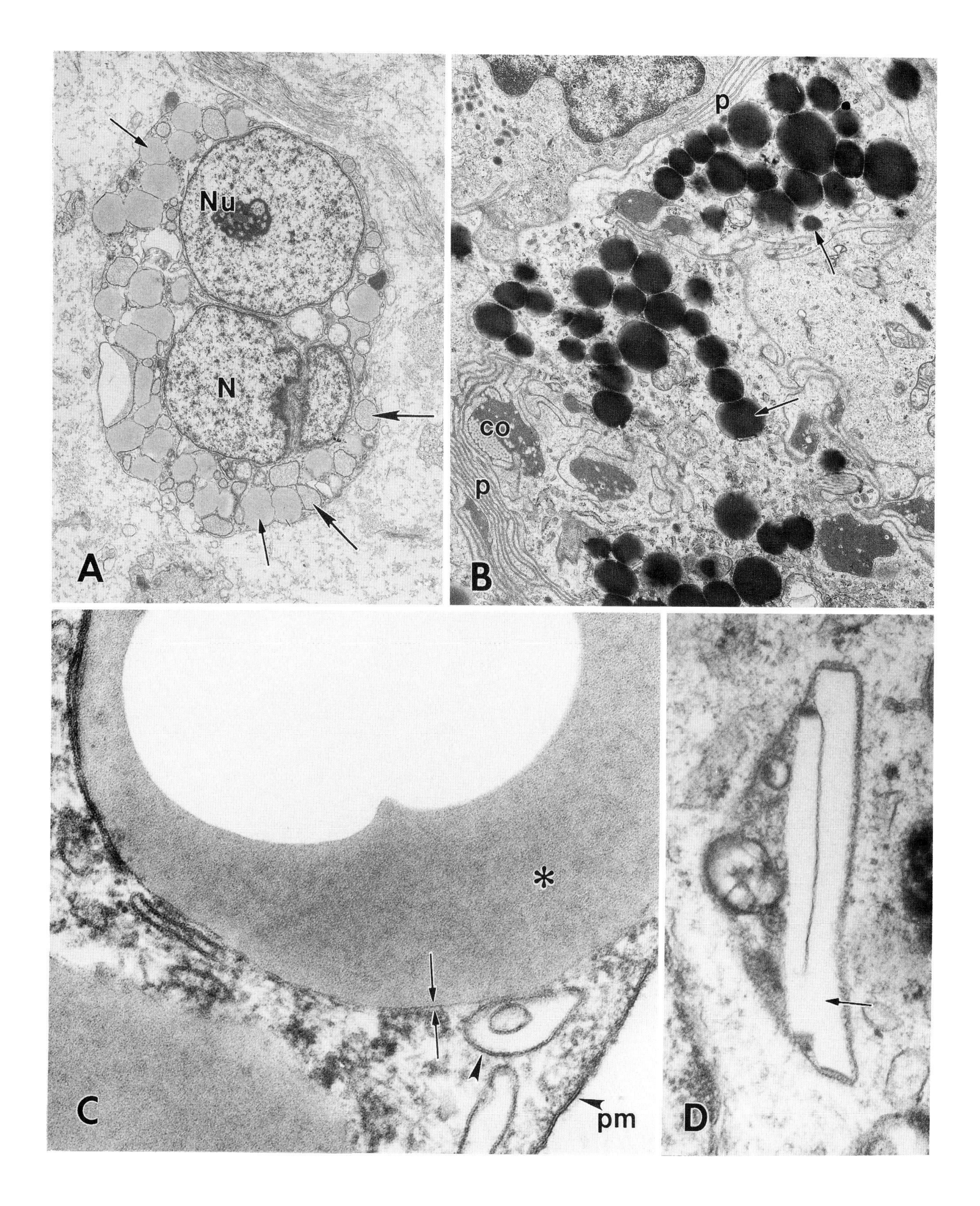
Nu
N
A
p
co
p
B
*
pm
C
D

ULTRASTRUCTURAL VARIATIONS; LAMELLAR LIPID

Lipid varies in appearance. There may be peripheral or central dense areas (Plates 42C and 43A), multiple small and irregular clear zones, or a lamellar organization. The lamellae may appear as single dense lines or trilaminar. The trilaminar lamellae resemble, but are more sharply defined than, true membrane (Tome et al., 1977).

Lipid droplets may show partial or complete lamellar transformation producing parallel or concentric lamellae (Plate 43); here, the purely descriptive term **lamellar body** or **concentric lamellar body** is appropriate. Of these, there are three main types, differing in nature and origin: pulmonary surfactant, myelin figures, and lysosomal lamellar bodies.

PULMONARY SURFACTANT

Cytoplasmic pulmonary surfactant characterizes type II pneumocytes and their tumors: alveolar cell adenomas/adenocarcinomas, bronchioloalveolar carcinomas, some peripheral pulmonary adenocarcinomas, and sclerosing hemangioma of lung.

SURFACTANT ULTRASTRUCTURE

Cytoplasmic surfactant has concentric lamellar organization (Plate 44A and 44B). Surfactant *outside* the cell similarly consists of rounded lamellar bodies, but material referred to as **tubular myelin** with a cross-hatching appearance may also be present (Williams, 1977; Kishikawa, 1990). Often, intracytoplasmic surfactant bodies appear to have a discrete limiting membrane and sometimes therefore seem to be lying in a vacuole (Plate 44A and 44B). At low magnification, surfactant bodies often appear dense and the individual lamellae are not resolved (Noguchi et al., 1986). Vesicles may be admixed with lamellae, and sometimes lamellae are straight and parallel rather than concentric (Ghadially et al., 1985).

MYELIN FIGURES

Concentric lamellar bodies similar to surfactant inclusions can be encountered in cells which are not of type II pneumocyte differentiation. The ensemble of clinical, histological, and immunocytochemical data, as well as whether the lamellar bodies are bounded by a clearly defined membrane, bear on whether lamellar bodies should be interpreted as surfactant or nonsurfactant lipid. Many of the nonspecific concentric lamellar bodies are probably just lipid having undergone lamellar transformation (e.g., Hull and Eble, 1988). Such bodies can also be referred to as **myelin figures, myelinosomes**, or **myelinoid bodies**. Some of these, however, are probably **myelin figure artifacts**—a term reflecting the possibility that they are formed as lipid leaflets are released during or before aldehyde fixation and then become fixed and stained during osmication (Glauert, 1974). They are usually found solitarily, either in the cytoplasmic matrix or extracellular space (Plate 44C); and they are often attached to lipid or found on or in mitochondria (Plate 44D). Some authors have considered extracellular examples as equivalent to the myelin of nerve sheath and have, therefore, regarded them as indicators of Schwann cell differentiation. They are, however, also found in a variety of tumors of unambiguously non-schwannian type (e.g., lymphomas), and they should be used diagnostically with caution.

LYSOSOMAL LAMELLAR BODIES

Lipid is commonly found in lysosomes (Chapter 17), and this lipid can be amorphous or of lamellar organiz-

Plate 43. **A**: Lipid droplets with a variety of appearances. Some have central and more irregular clear areas representing lipolysis (*; see also Plate 42C). Others show more or less aligned lamellae (*arrows*). Lipid-rich renal cell carcinoma metastatic to supraclavicular node. ×8400. **B**: Lipid droplets in varying states of concentric lamellar transformation: 1 is completely amorphous; 2 shows predominantly *peripheral* lamella formation (the individual lamellae are not resolved at this magnification); 3 shows complete concentric lamellar transformation. ×8800. **C**: Detail from **B** showing individual lamellae. Some are tangentially sectioned (*), and only those sectioned at right angles (*arrows*) are imaged sharply. **B** and **C** are from deparaffinized tissue, and the sharp delineation of lamellae in **C** may in part be due to the use of *en bloc* uranyl acetate during tissue processing (Fehrenbach et al., 1995). Clear-cell neuroendocrine carcinoma, pancreas. ×95,200. N, nucleus; Nu, nucleolus.

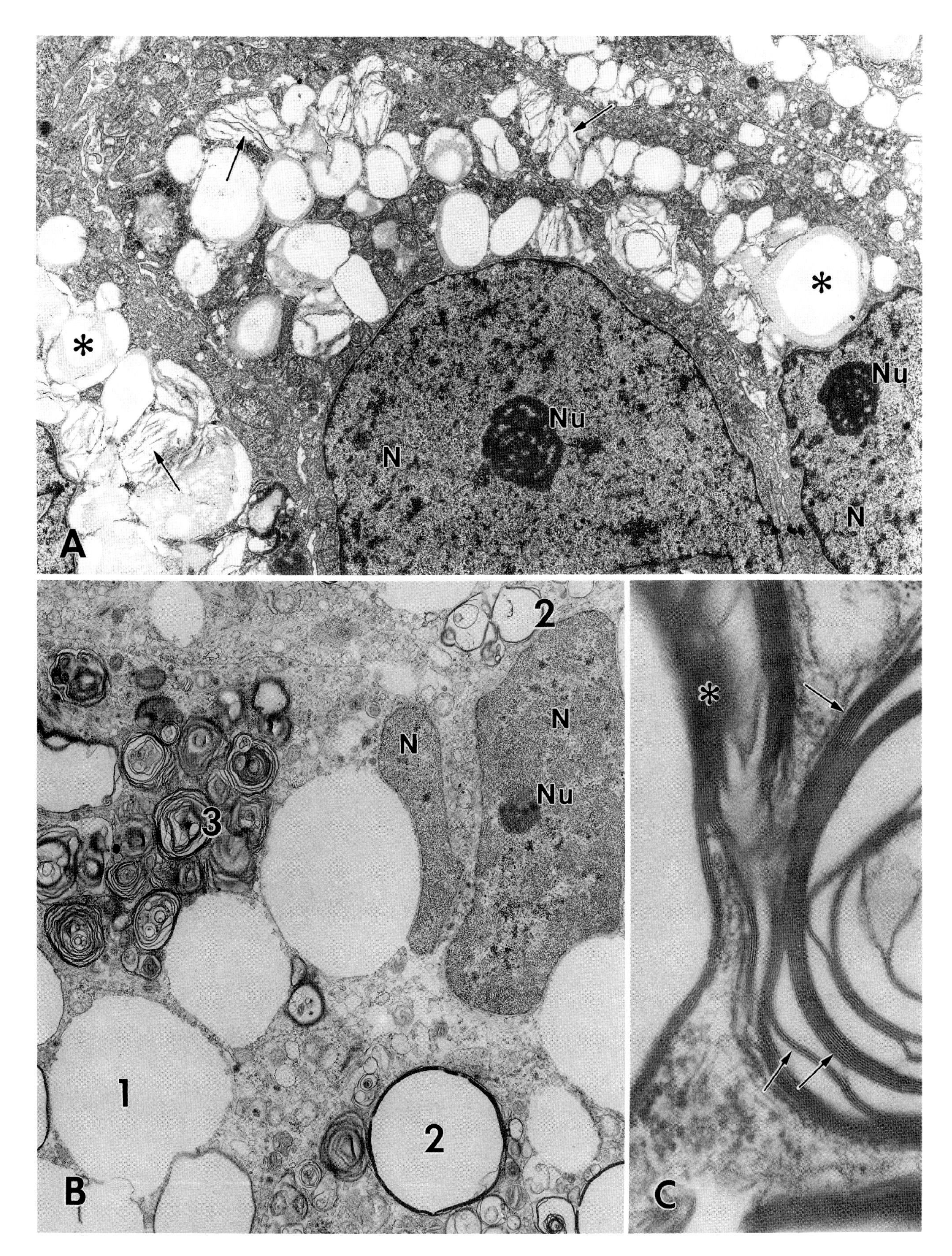
A
*
*
Nu
Nu
N
N
B
2
N
N
Nu
3
1
2
C
*

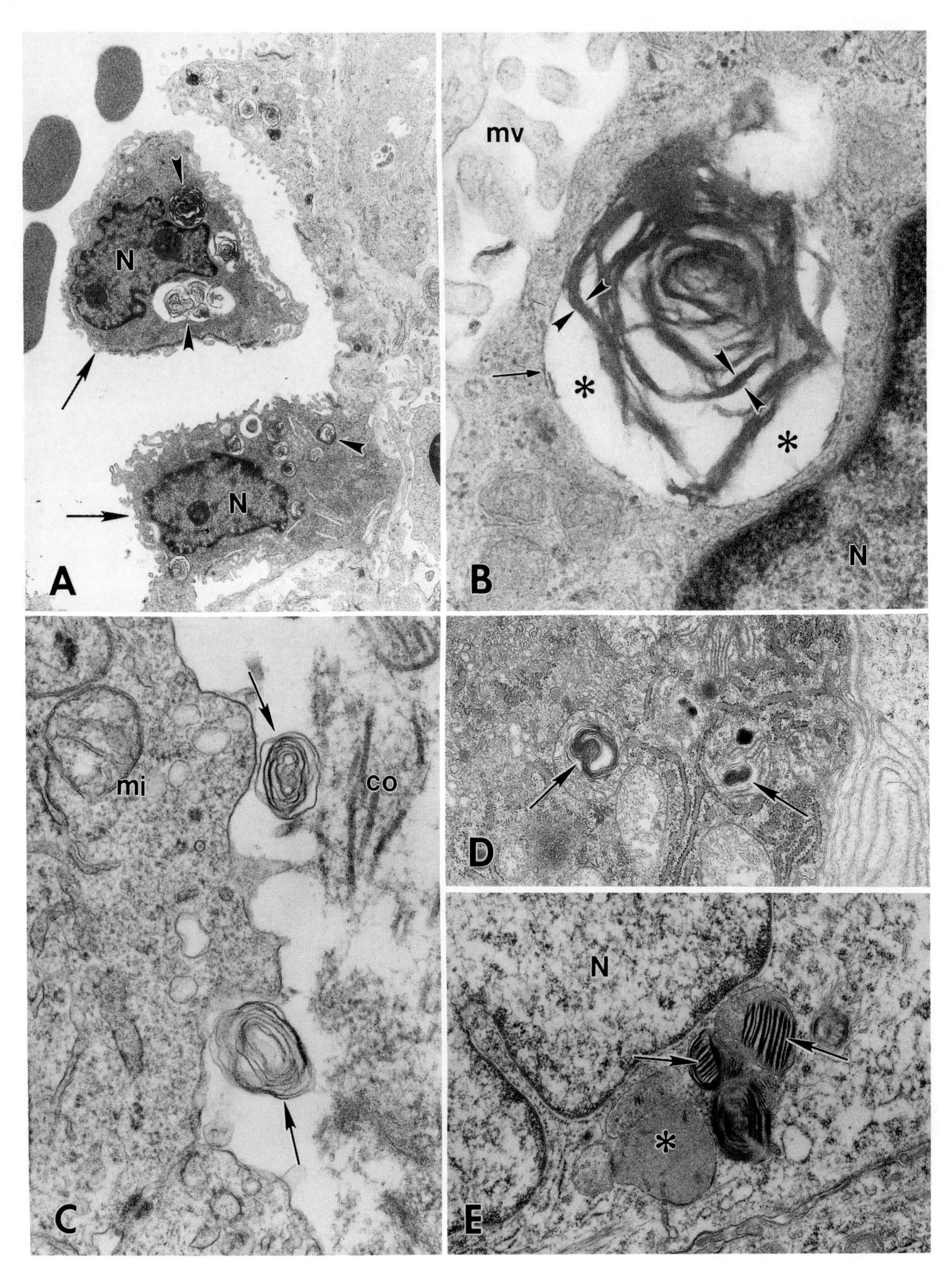
N
N
A
mv
B
N
mi
co
C
D
N
E

ation. The most dramatic examples of lamellar body formation in lysosomes is found in the non-neoplastic lysosomal storage diseases (Resibois et al., 1970; Martin and Ceuterick, 1988; Elleder, 1989), but such bodies are also found in tumor cells (Plate 44E).

The **membrane-coating granules** (also called **keratinosomes** and **Odland bodies**) of keratinizing squamous cells also have a lamellar content (El-Labban and Wood, 1982; Ferey et al., 1985).

Plate 44. **A**: Type II pneumocytes (*arrows*) in human lung containing light and dark surfactant bodies (*arrowheads*). Note similarity to nonsurfactant lamellar lipid in Plate 43B. ×4600. **B**: Detail from **A**, showing concentric masses of lamellae (*arrowheads*) which individually are not resolved. There is also an apparent limiting membrane (*arrow*), although one cannot rule out the possibility that this is the outermost lamella and that the space (*) represents retraction artifact. ×53,700. **C**: Extracellular myelin figure artifacts (*arrows*). Note mitochondrion for size. Epithelioid malignant schwannoma, thigh. ×39,000. **D**: Myelin figure artifacts associated with mitochondria (*arrows*) in a reactive plasma cell. Langerhans cell granulomatosis, submental lymph node. ×14,600. **E**: Lamellar lipid (*arrows*) in secondary lysosomes. One lysosome (*) is free of lipid. Epithelial mesothelioma, peritoneum. ×19,600. co, collagen; mi, mitochondrion; mv, microvilli; N, nucleus; Nu, nucleolus.

20

Glycogen

Glycogen is the principal polysaccharide reserve material in mammalian cells. It is normally found in the cytoplasmic matrix between organelles, more rarely within organelles such as lysosomes.

Small amounts of glycogen are almost ubiquitous in cells, but it is abundant and therefore diagnostically significant in muscle and liver cells and their tumors, some clear-cell carcinomas and sarcomas, and Ewing's tumor. It is important to recognize the range of ultrastructural features of glycogen given its tendency to be leached out during processing and therefore to give rise to clear cytoplasmic spaces (Plate 47).

ULTRASTRUCTURE OF GLYCOGEN

Glycogen is found in two structural forms. Small particles about 30 nm across are **monoparticulate** or **beta-glycogen**. They can be found in extensive lakes (Plate 45A), in smaller foci, or in small numbers dispersed amongst other cell constituents (Plate 45C). Larger aggregates of such small particles are **alpha-rosettes** (Plate 46A), which can measure up to ~200 nm across. Alpha-glycogen can also form extensive lakes throughout the cytoplasm; here, high magnification sometimes reveals a mixture of alpha-rosettes and smaller less-well-defined particles (Plate 46B).

The density of glycogen varies considerably from pale-staining to extremely electron-dense (Plate 46). Lakes of glycogen may be infiltrated with smooth endoplasmic reticulum (Wills, 1992) and lipid (Plate 46B).

Glycogen can be leached out by certain rinsing procedures, especially *en bloc* uranyl acetate. This results in partially or completely extracted glycogen pools (Plate 47A), sometimes bordered by or containing nondescript dense granules (Plate 47B and 47C); these are presumably of glycogen, but they do not conform to either alpha or beta glycogen morphologically.

Plate 45. **A**: Perinuclear glycogen lake. Adenocarcinoma of unknown origin metastatic to parotid. ×9000. **B**: Detail of beta-glycogen granules from a glycogen lake. Rhabdomyosarcoma, inferior vena cava. From a block supplied by Dr. Y. Kuwashima (Saitama, Japan). ×60,800. **C**: Dispersed beta-glycogen particles (*large arrows*). Note rounded contour, more homogeneous density, and larger size (30 nm) compared with ribosomes (20 nm) (*small arrows*). Note also microtubules, as well as myofilaments with focal densities. Desmoplastic myofibroblast from infiltrating duct carcinoma. ×199,000. d, desmosome; fd, focal density; g, glycogen; mt, microtubules; my, myofilaments.

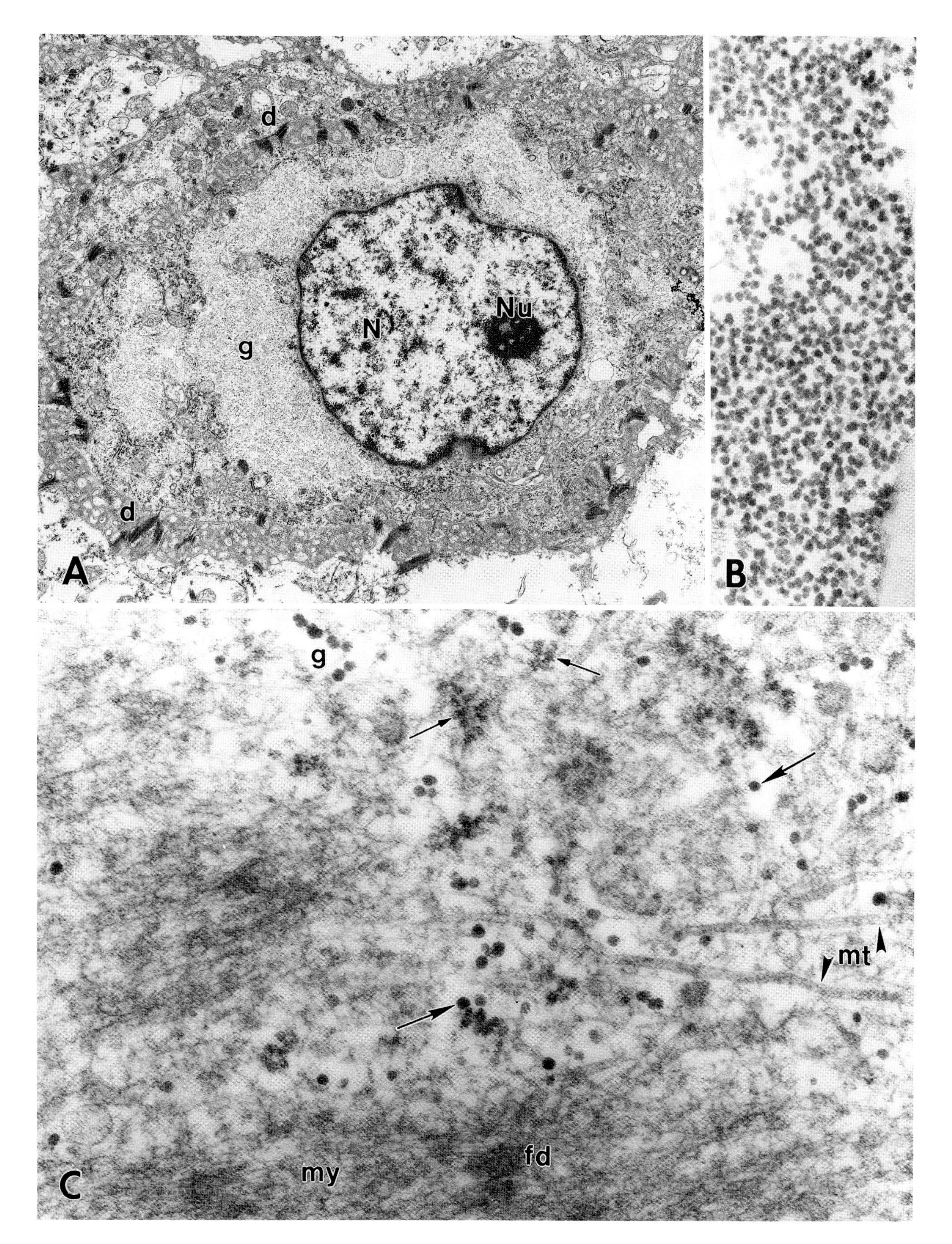
d
N
Nu
g
d
A
B
g
mt
my
fd
C

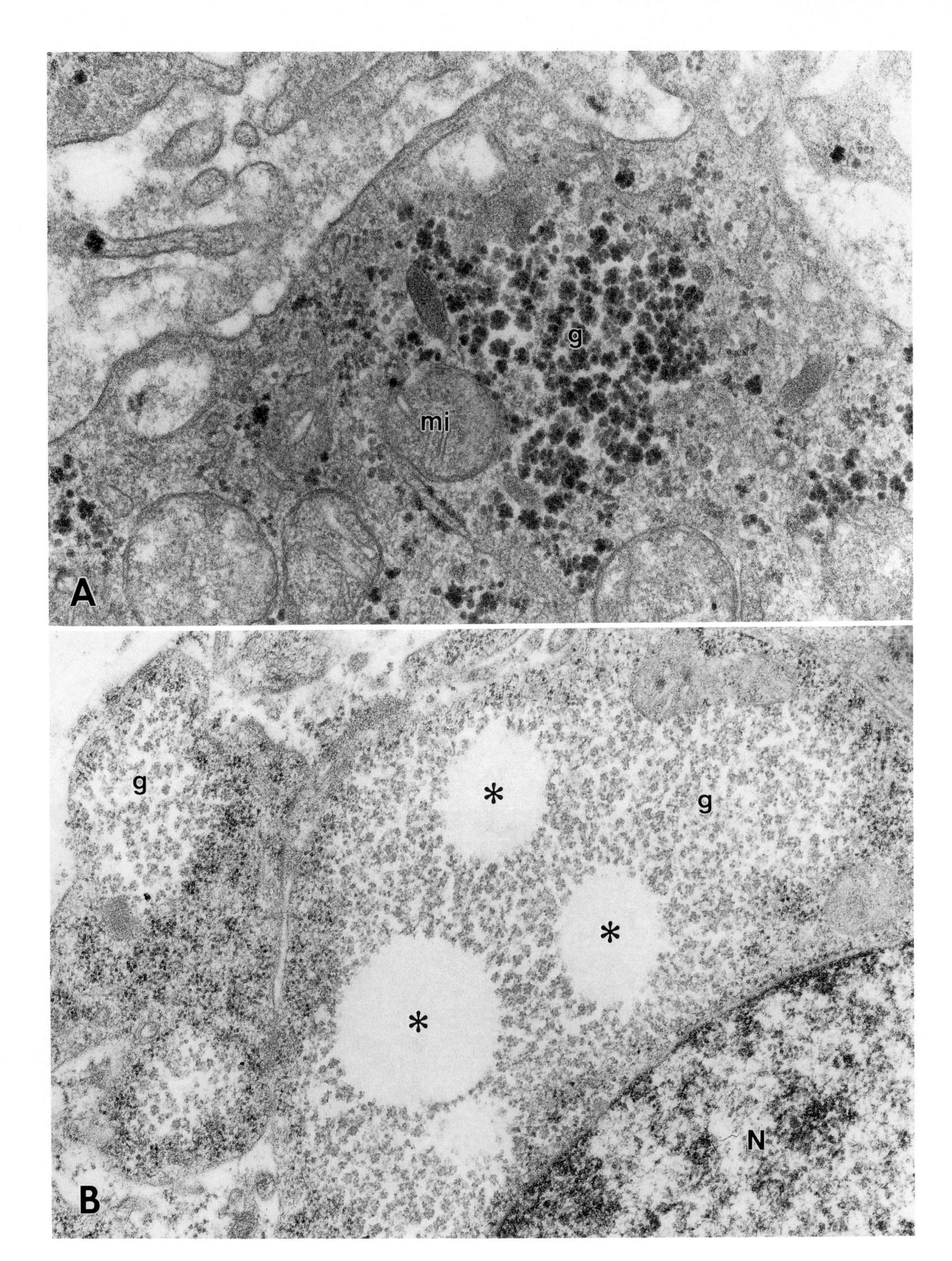
g
mi
A
g
*
g
*
*
N
B

Plate 46. **A**: Focus of dense alpha-rosettes. Glycogen-rich metastatic malignant melanoma, supratrochlear node. ×61,100. **B**: Lipid droplets (*) in an area of pale-staining glycogen particles. Ewing's tumor, subcutis of thigh. ×51,000. g, glycogen; mi, mitochondrion; N, nucleus.

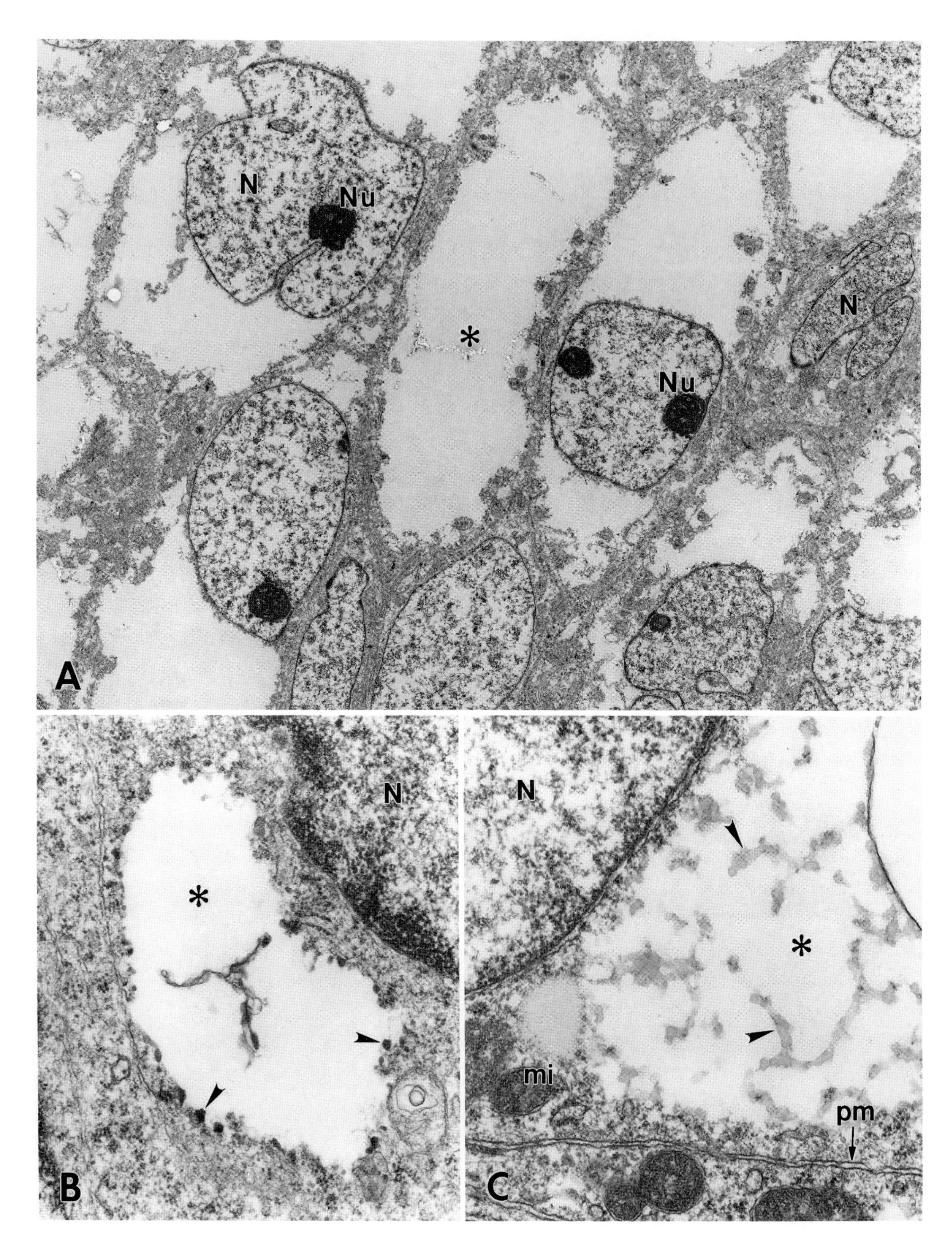
N
Nu
*
N
Nu
A
N
*
B
N
*
mi
pm
C

Plate 47. **A**: Completely leached-out glycogen (*) from glycogen-rich clear-cell carcinoma, breast. × 5300. **B**: Small focus of leached out glycogen (*) with peripheral residual granules (*arrowheads*). Peripheral primitive neuroectodermal tumor, dermis/subcutis of scalp. × 35,000. **C**: Reticulate residual material (*arrowheads*) from a glycogen lake (*). Extraskeletal Ewing's tumor, inguinal soft tissues. × 38,200. **A–C**: After uranyl acetate *en bloc* staining. mi, mitochondrion; N, nucleus; Nu, nucleolus; pm, plasma membrane.

21

Peroxisomes

Peroxisomes (also called **microbodies**) contain oxidative enzymes (e.g., catalase) which use oxygen in the respiratory breakdown of substrates. Hydrogen peroxide produced in these reactions is decomposed by catalase, the most abundant oxidative enzyme in peroxisomes.

DISTRIBUTION AND ULTRASTRUCTURE OF PEROXISOMES

Peroxisomes are ubiquitous in cells but have mainly been reported in non-neoplastic human liver and kidney, hepatocellular carcinomas, and a hepatoblastoma cell line. It is likely that they have been underidentified in human tumors.

Peroxisomes are more or less rounded single-membrane-bound granules with a dense matrix (Plate 48), and sometimes a crystalloid is seen (Plate 48A). They have a wide size range: 100 nm to ~1 μm. The larger peroxisomes are typically found in hepatocytes. Usually, peroxisomes lack the halo typical of neuroendocrine granules and lysosomes, and often there is a patchiness to the granular matrix (Plate 48C). On occasion also, one may see tail-like extensions of the limiting membrane (Plate 48C and 48D). Although originally identified by the purely ultrastructural criteria of electron density and sometimes an internal crystalloid, they are now best demonstrated by catalase enzyme cytochemistry (Plate 48D).

Plate 48. **A**: Large dense round-to-oval peroxisomes in clusters (*arrows*), one with a crystalloid (*arrowhead*). Note pale-staining poorly preserved mitochondria, and also note cleft-like expansions of rough endoplasmic reticulum. Fibrolamellar hepatocellular carcinoma. ×20,000. [Micrograph courtesy of Dr. T. Caballero (Granada, Spain). Reproduced from Caballero et al. (1985), with permission from Blackwell Science, Ltd.] **B**: Tumor cell containing several small dense granules (*arrows*). Tumor diagnosed as neuroendocrine carcinoma (retroperitoneum) by light microscopy and immunostaining. ×13,300. **C**: Detail of granule (*) from **B**. It has a membranous tail or is in close contact with an element of smooth endoplasmic reticulum (*arrowhead*). ×61,100. **D**: Peroxisomes stained for catalase and showing membranous tails (*arrowheads*). Non-neoplastic liver in patient with giant-cell carcinoma of bronchus. ×21,000. [Micrograph courtesy of Dr. De Craemer (Brussels). Reproduced from De Craemer et al. (1993), with permission from J. B. Lippincott Company.] g, glycogen; mi, mitochondria; N, nucleus; Nu, nucleolus; rER, rough endoplasmic reticulum.

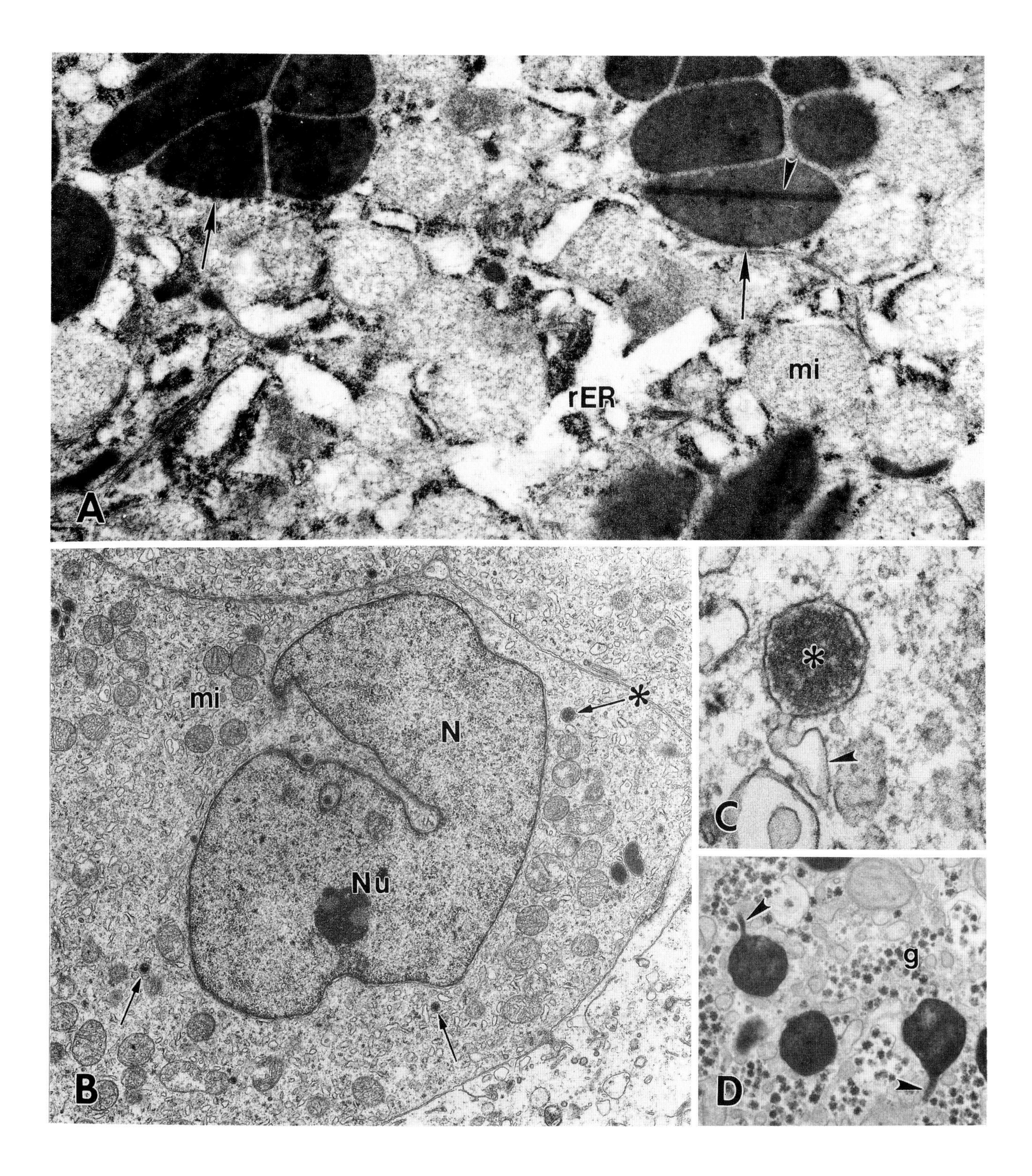
A
rER
mi
B
mi
N
Nu
*
C
*
D
g

22

Contractile Filaments

The contractile filaments, **actin** and **myosin**, are ubiquitous in cells. They constitute one component of the **cytoskeleton**, the complex of filamentous proteins found mainly in the cytoplasm between organelles. The cytoskeleton functions in the development and maintenance of cell shape; it also functions in movement, either of organelles, parts of cells, or whole cells. The other major cytoskeletal components are **intermediate filaments** (Chapter 24) and **microtubules** (Chapter 25).

Actin and myosin filaments are collectively referred to as **myofilaments** because, when present in large numbers and in certain patterns of organization, they characterize muscle cells. The term *microfilament* for actin filament is encountered in the literature but is obsolete.

SMOOTH-MUSCLE MYOFILAMENTS

Smooth-muscle myofilaments constitute one of the most important ultrastructural markers of smooth-muscle cells. However, they exist in other cells which have features allowing them to be considered as not primarily or not purely smooth muscle in nature (e.g., myoepithelium, pericytes, myofibroblasts).

- Smooth-muscle myofilaments are most commonly found in: leiomyomas, leiomyosarcomas, myoepitheliomas, hemangiopericytomas; myofibroblastic lesions such as nodular fasciitis, myofibromatoses, fibroma of tendon sheath, infantile digital fibromas; epithelial tumors (Plate 52B) such as invasive squamous carcinomas, spindle cell carcinomas, pancreatic serous cystadenomas, lacrymal gland choristomas, adenoid cystic carcinomas.
- The following lesions (not primarily myogenic, pericytic, myofibroblastic, or myoepithelial) contain modestly developed smooth-muscle myofilaments: granulosa cell tumors, osteosarcomas, desmoplastic malignant melanomas, and xanthogranulomas.
- Smooth-muscle myofilaments are seen in some *non-neoplastic* endothelium and epithelium (e.g., lens epithelium in anterior capsular cataract, atrophic renal tubules) and in reactive subserosal cells.

Plate 49. **A**: Smooth-muscle cell with abundant myofilaments and a few organelles at each end of the nucleus (*). Note focal densities (*arrowheads*) and attachment plaques (*arrows*). Normal myometrium. ×8500. (Reproduced from Dardick I. *Handbook of Diagnostic Electron Microscopy for Pathologists-in-Training*, New York, Igaku-Shoin, 1996, with permission.) **B**: Detail of myofilaments and focal densities. Note thin actin (*small arrow*) and thicker myosin filaments (*large arrows*), lysosome, and extracellular space. Uterine leiomyoma. ×92,500. ecm, extracellular matrix; fd, focal density; ly, lysosome; my, myofilaments; N, nucleus.

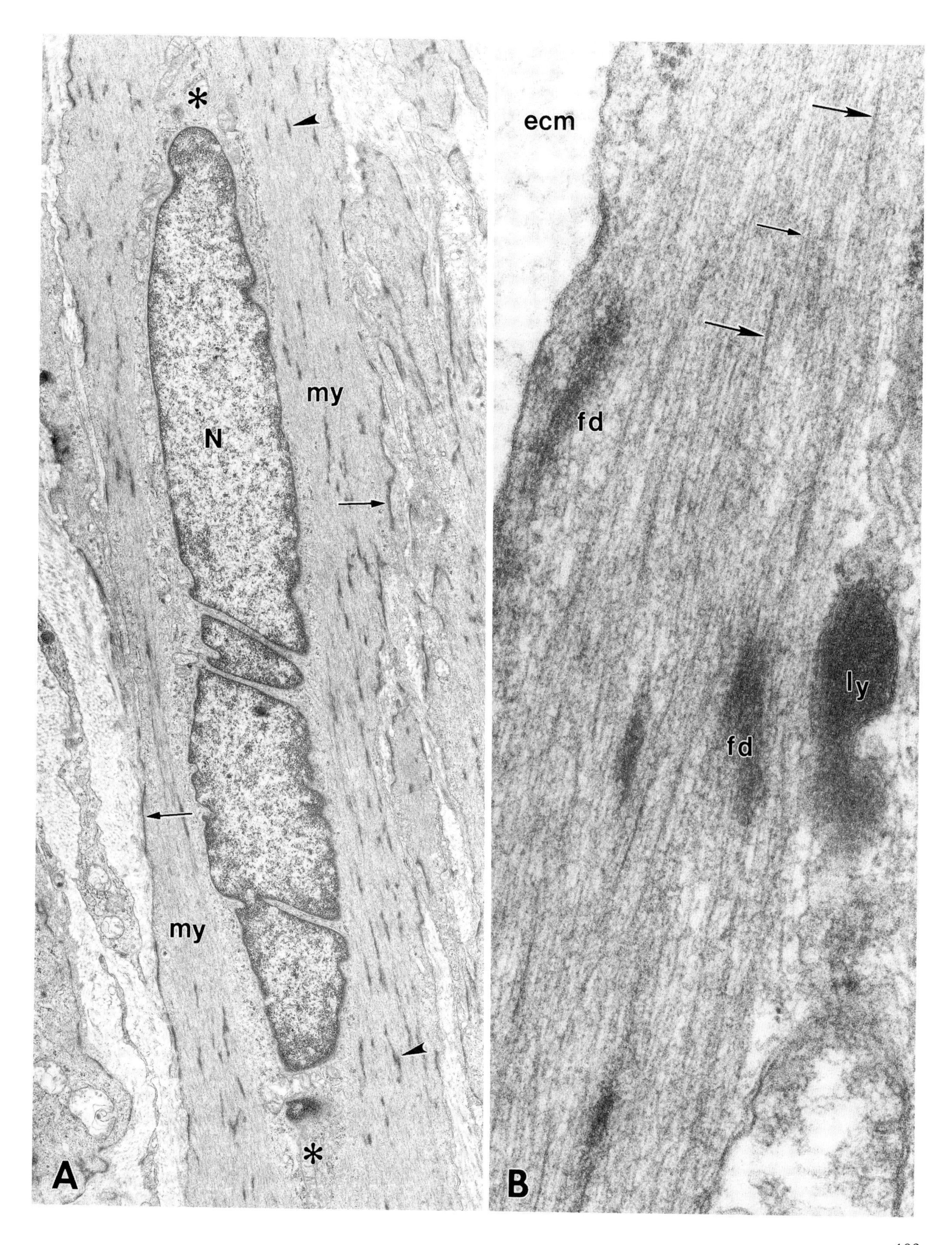
*
my
N
my
*
A
ecm
fd
fd
ly
B

ULTRASTRUCTURE AND ARRANGEMENT OF SMOOTH-MUSCLE MYOFILAMENTS

Smooth-muscle myofilaments are usually arranged in the long axis of the cell (Plate 49A). When especially abundant they may leave only a small paranuclear focus of organelles. At low magnification they may give a pale-staining effect to the cytoplasm (Plate 49A) which only higher magnification reveals as filamentous (Plate 49B). In myofibroblasts and some myoepithelial cells, myofilaments are fewer and are often organized as bundles under the cell surface membrane (Plate 69).

In standard transmission electron microscopy preparations, actin filaments predominate (Plate 49B). These are straight, very fine, and ~5–7 nm thick. Myosin filaments, which are also straight but thicker (~15 nm), are present in smaller numbers (Plate 49B). Interspersed amongst these filaments are localized areas of dense material (Plate 49A and 49B), believed to be sites at which contractile and intermediate filaments interact. It is preferable to call these **focal densities**; they are, however, commonly known as *dense bodies.* Focal densities are typically 200–400 nm long and lie in the long axis of the myofilaments, but they can vary in size and shape (Plate 50A and 50B).

Tracts of fine filaments *without* dense bodies can be seen in many nonmuscle cells. Therefore, thin filaments must be accompanied by focal densities to be considered as a marker of smooth-muscle differentiation.

Smooth-muscle cells typically possess foci of dense material on the underside of the plasma membrane (Plate 50B–D). These are considered as membrane-attached versions of the focal density and appear to be sites of contact between the cell membrane and the peripheral cytoskeleton. They are best described as **attachment plaques**, in preference to *subplasmalemmal densities* and *membrane thickenings.* In vascular smooth-muscle cells, attachment plaques can be exceptionally large (Plate 50C). Focal densities, attachment plaques, and myofilaments survive paraffin embedding.

VARIANTS OF FOCAL DENSITIES AND ACTIN FILAMENT AGGREGATES

Occasionally, structures resembling focal densities but with almost no discernible associated myofilaments can be found either embedded in intermediate filament masses (Plate 51A) or lying free in peripheral cytoplasm (Plate 51B and 51C). One distinctive example is the inclusion body found in digital fibroma and other similar lesions (Plate 52A). Generally, these are tentatively interpreted as **solitary focal densities** or **compacted actin filaments**: some are fusiform, they are frequently subplasmalemmal in location, some have an actin-filament-like substructure, and some (e.g., the inclusion body found in digital fibroma) stain for smooth-muscle actin.

STRIATED-MUSCLE MYOFILAMENTS

The contractile filaments of skeletal and cardiac muscle are referred to as **striated-muscle** or **sarcomeric myofilaments**.

Sarcomeric myofilaments are found in: rhabdomyomas, rhabdomyosarcomas, and rhabdomyoblastic cells present as sarcomatous, metaplastic, or hamartomatous elements in tumors such as malignant mixed mesodermal tumor, carcinosarcomas, malignant mesenchymomas, Triton tumors, hepatoblastomas, and Wilms' tumor [see Agamanolis et al. (1986) for further examples].

ULTRASTRUCTURE AND ARRANGEMENT OF SARCOMERIC MYOFILAMENTS

Sarcomeric myofilaments form bundles (**myofibrils**) (Plate 53A) in which thick myosin and fine actin filaments lie in parallel arrays (Plate 53B), interacting with

Plate 50. **A**: Two rather broad focal densities (*). Orientated filaments are not easy to see, but the material in which the focal densities are embedded is almost certainly smooth-muscle actin on the basis of overall ultrastructure of the cell and immunostaining in this specimen. Stromal cell, Wharton's jelly of umbilical cord. ×42,200. **B**: A large focal density (*) mostly at right angles to the direction of the myofilaments, almost resembling a Z-disk (see sarcomeric myofilaments, below). The structure, however, is in a histologically typical leiomyosarcoma (soft tissues, thigh), positive for smooth-muscle actin. Note lamina and attachment plaques (*arrowheads*). ×42,100. **C**: Large membrane-bound focal densities in vascular smooth-muscle cells (*arrowheads*). Monocytoid B-cell lymphoma, cervical node. ×13,600. **D**: Extended attachment plaques (*arrowhead*), myofilaments, and focal density. Observe differences between myofilaments and intermediate filaments, and note lamina and extracellular space (*). This seems to contain ribosome-like particles, presumably released from nearby disrupted cells. Uterine leiomyoma. ×70,700. [Reproduced from Eyden et al. (1992), with permission from Springer-Verlag.] ecm, extracellular matrix; fd, focal density; hc, heterochromatin; if, intermediate filaments; la, lamina; my, myofilaments.

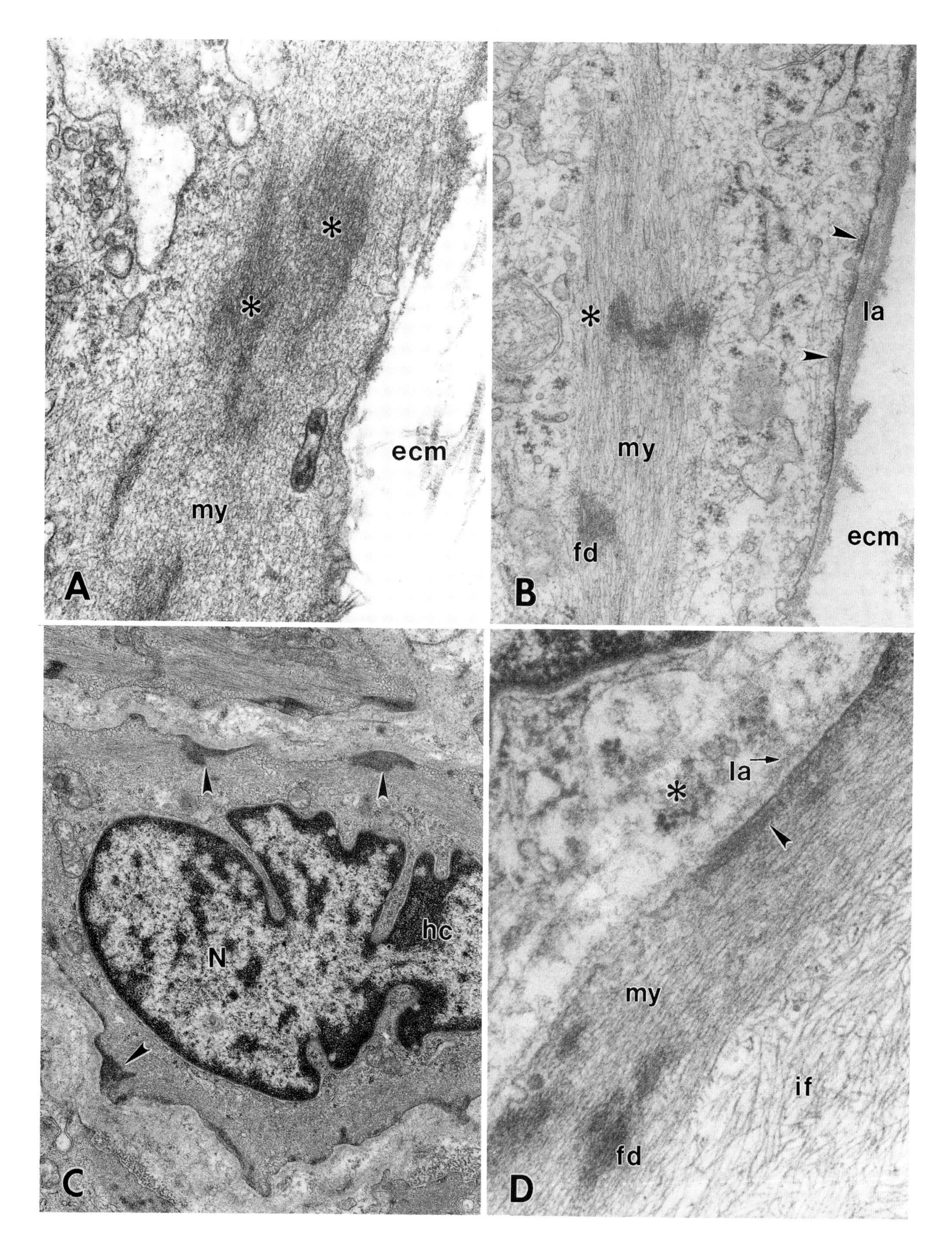
A
*
*
ecm
my
B
*
la
my
fd
ecm
C
N
hc
D
la
*
my
if
fd

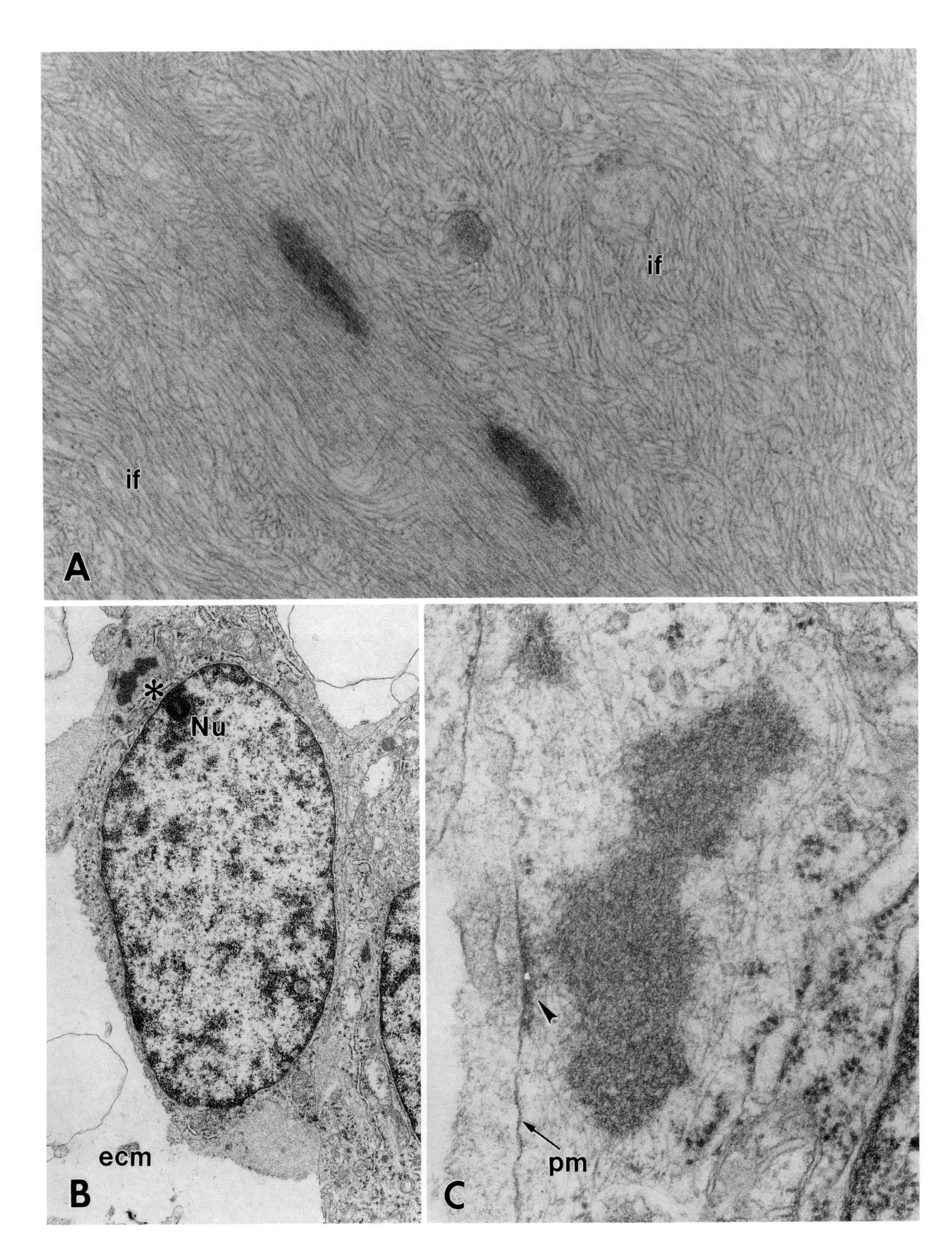
if
if
A
*
Nu
ecm
B
pm
C

foci of dense material traversing the filament bundle at right angles. These are **Z-disks** into which the actin filaments insert. In cross section (Plate 53C), the filaments are geometrically arranged, often with a hexagonal symmetry. In rhabdomyomas and well-differentiated rhabdomyosarcomas, this structural order is often retained (Plate 54); on the other hand, rhabdomyosarcomas can show a wide range of myofilament bundle disorganization. In comparatively poorly differentiated rhabdomyosarcomas, cross-striations by phosphotungstic acid–hematoxylin staining may be absent through poor organization of myofibrils and Z-disks. At low power, rhabdomyoblastic cells are sometimes seen to contain amorphous masses of dense material where myofibrils and Z-disks in particular appear to be absent (Plate 55A). High magnification, however, reveals myosin and actin filaments in varying states of organization (Plate 55B). Since clearly defined longitudinal myofilament arrays may be absent and since in some instances smooth-muscle focal densities and sarcomeric Z-disks may resemble one another (Plate 50B), good evidence for sarcomeric differentiation can be provided by seeing hexagonal arrays of thick and thin filaments in cross section (Plate 53C). In well-developed sarcomeric myofibrils, however, it is also possible to see actin alone or myosin alone, in either hexagonal or quadratic arrays (Plate 55C).

Z-disks may sometimes look more solid and block-like than plate-like, and they may also reveal transverse striations at right angles to the long axis of the sarcomere (reminiscent of nemaline bodies; see Chapter 23). In some myofibrils, myosin filaments may be inconspicuous; by contrast, in **ribosome–myosin complexes**—regarded by some as the minimal ultrastructural criterion for sarcomeric differentiation (Plate 55D)—actin appears to be absent or poorly defined.

Plate 51. **A**: Structures resembling focal densities with few associated myofilaments embedded in intermediate filaments. Uterine leiomyoma. × 70,700. **B** and **C**: Peripherally located dense bodies (*) possibly representing solitary focal densities or compacted actin filaments. Recurrent parapharyngeal leiomyosarcoma. In **C**, note submembranous dense plaque of similar appearance (*arrowhead*). **B**, × 8800; **C**, × 70,700. ecm, extracellular matrix; if, intermediate filaments; Nu, nucleolus; pm, plasma membrane.

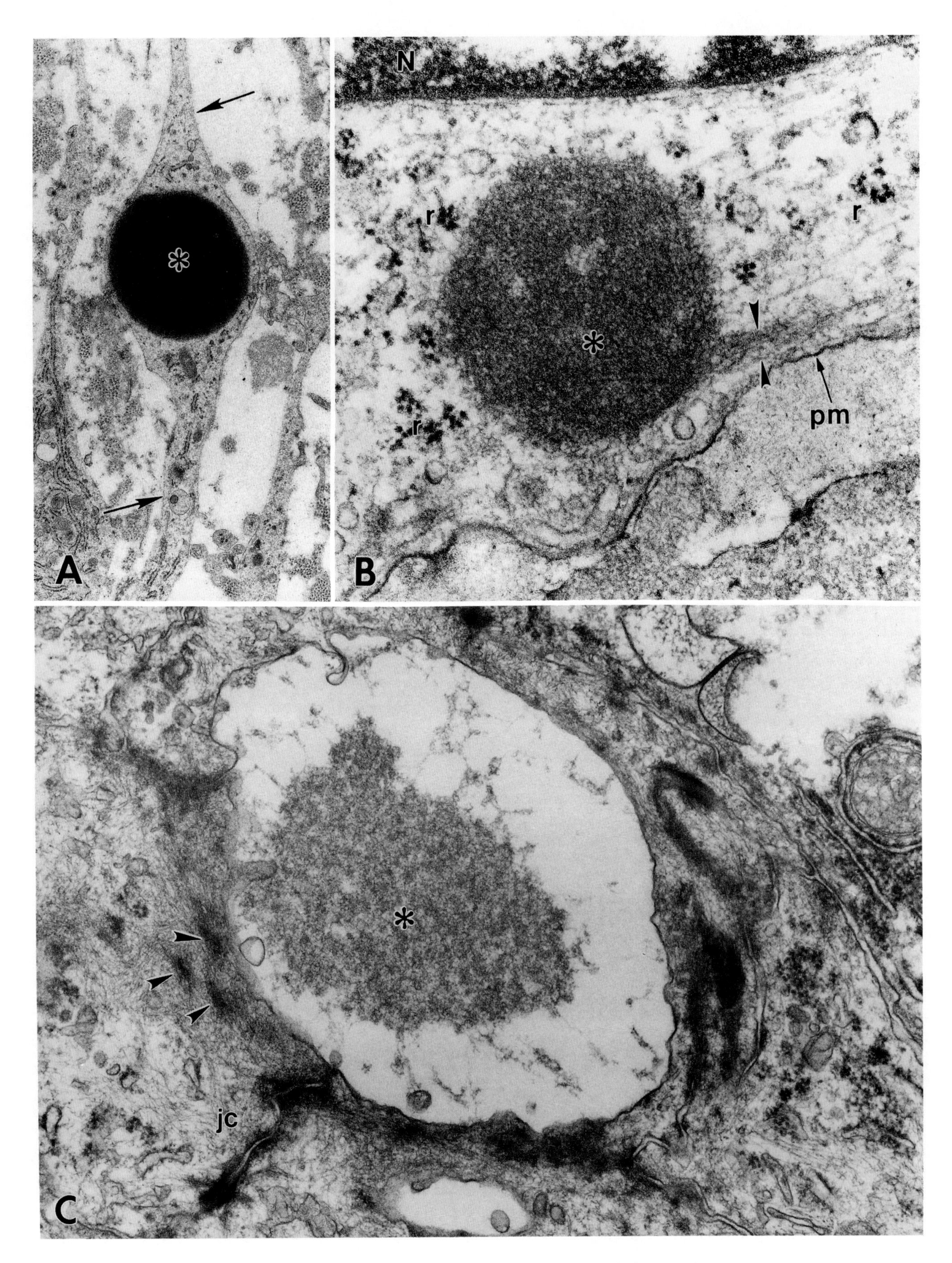
A
B
N
r
r
r
pm
C
jc

Plate 52. **A**: Rounded inclusion body. Myofibroblast (*arrows*) in stroma of phyllodes tumor, breast. From a block courtesy of Dr. N. Hiraoka (Tokyo). **B**: Detail of a similar inclusion to show finely textured filamentous substructure and association with tract of fine peripheral filaments (*arrowheads*). ×93,600. [Reproduced from Eyden et al. (1989), with permission from the *Journal of Submicroscopic Cytology and Pathology*.] **C**: Apical region of cytoplasm containing a terminal web of fine filamentous material with condensations reminiscent of focal densities (*arrowheads*). Note junctional complex and secretory material (*). Pleomorphic adenoma, parotid. ×39,600. jc, junctional complex; N, nucleus; pm, plasma membrane.

Z
Z
sER
A
C
mi
g
Z
mi
Z
B

Plate 53. **A**: Striated (cardiac) muscle showing more or less cohesive myofibrils made up of sarcomeres. Note the ~800-nm-diameter mitochondria for size, and also note the fine reticulum of smooth sarcoplasmic reticulum. x13,000. (Reproduced from Dardick I. *Handbook of Diagnostic Electron Microscopy for Pathologists-in-Training*, New York, Igaku-Shoin, 1996, with permission.) **B**: Detail of sarcomere in long section to show thick (*large arrows*) and thin (*small arrows*) filaments and Z-disk material. Note mitochondrion and monoparticulate (beta) glycogen. Human pectoral skeletal muscle. × 60,000. **C**: Sarcomeric myofibril in cross section to show ordered array of thick and thin filaments. Rhabdomyoblast in carcinosarcoma, urinary bladder. × 112,000. [Reproduced from Cross et al. (1989), with permission from Springer-Verlag; Germany); and from Dardick I. *Handbook of Diagnostic Electron Microscopy for Pathologists-in-Training*, New York, Igaku-Shoin, 1996, with permission.)] g, glycogen; mi, mitochondria; sER, sarcoplasmic reticulum; Z, Z-disk.

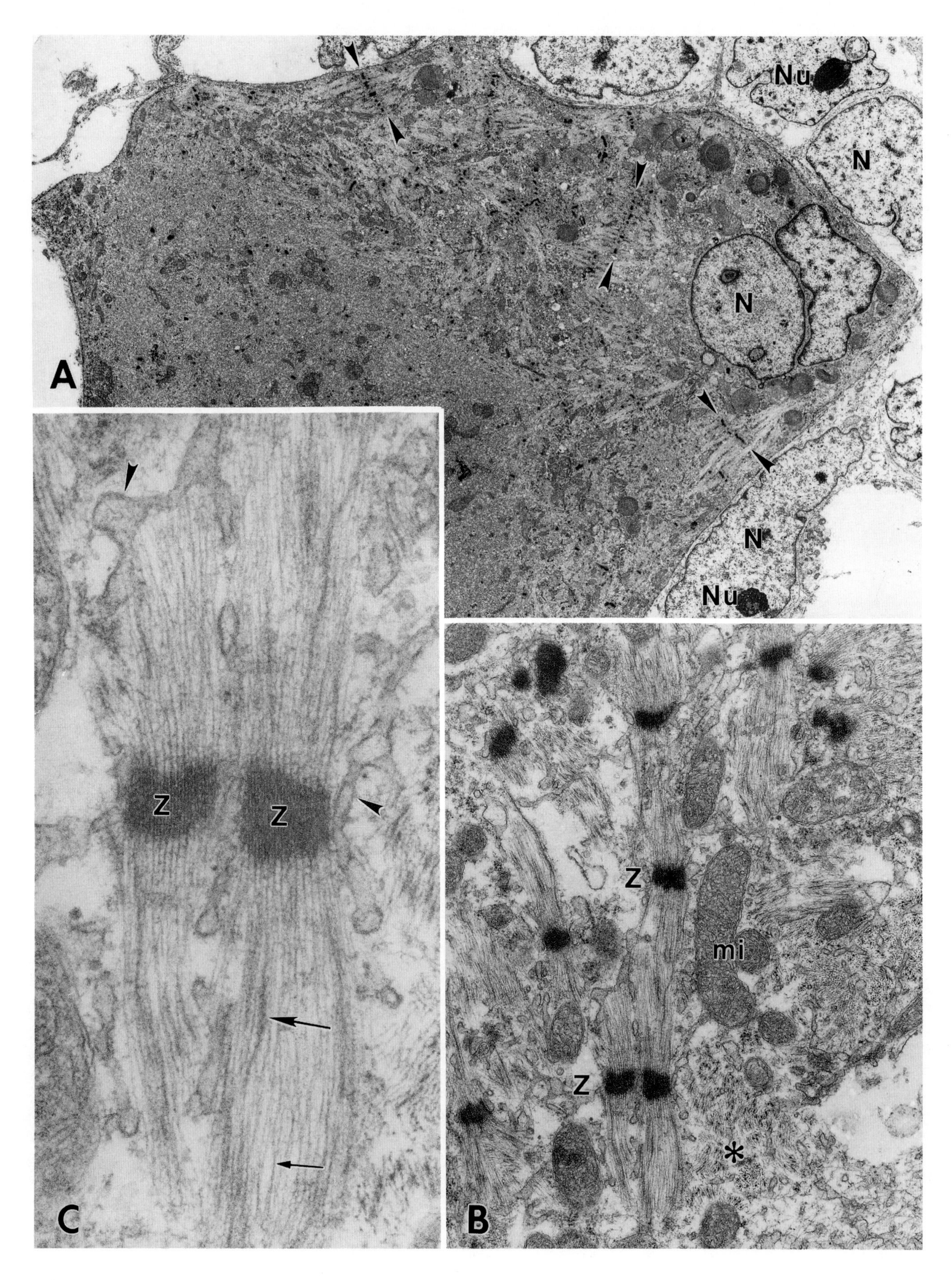
Nu
N
N
A
N
Nu
Z
Z
C
Z
mi
Z
*
B

Plate 54. **A**: Large rhabdomyoblast containing mitochondria and aligned Z-disks (*arrowheads*). Note nuclei of surrounding undifferentiated tumor cells. ×3200. **B**: Myofibrils with Z-disks. Asterisk (*) indicates less organized myofilaments. ×20,400. **C**: Detail of myofibril showing actin (*small arrow*) and myosin (*large arrow*) filaments. Note also sarcoplasmic reticulum (*arrowheads*) over surface of myofibril. ×69,000. **A–C**: Botryoid type of rhabdomyosarcoma, ethmoid sinus. mi, mitochondria; N, nuclcus; Nu, nucleolus; Z, Z-disk.

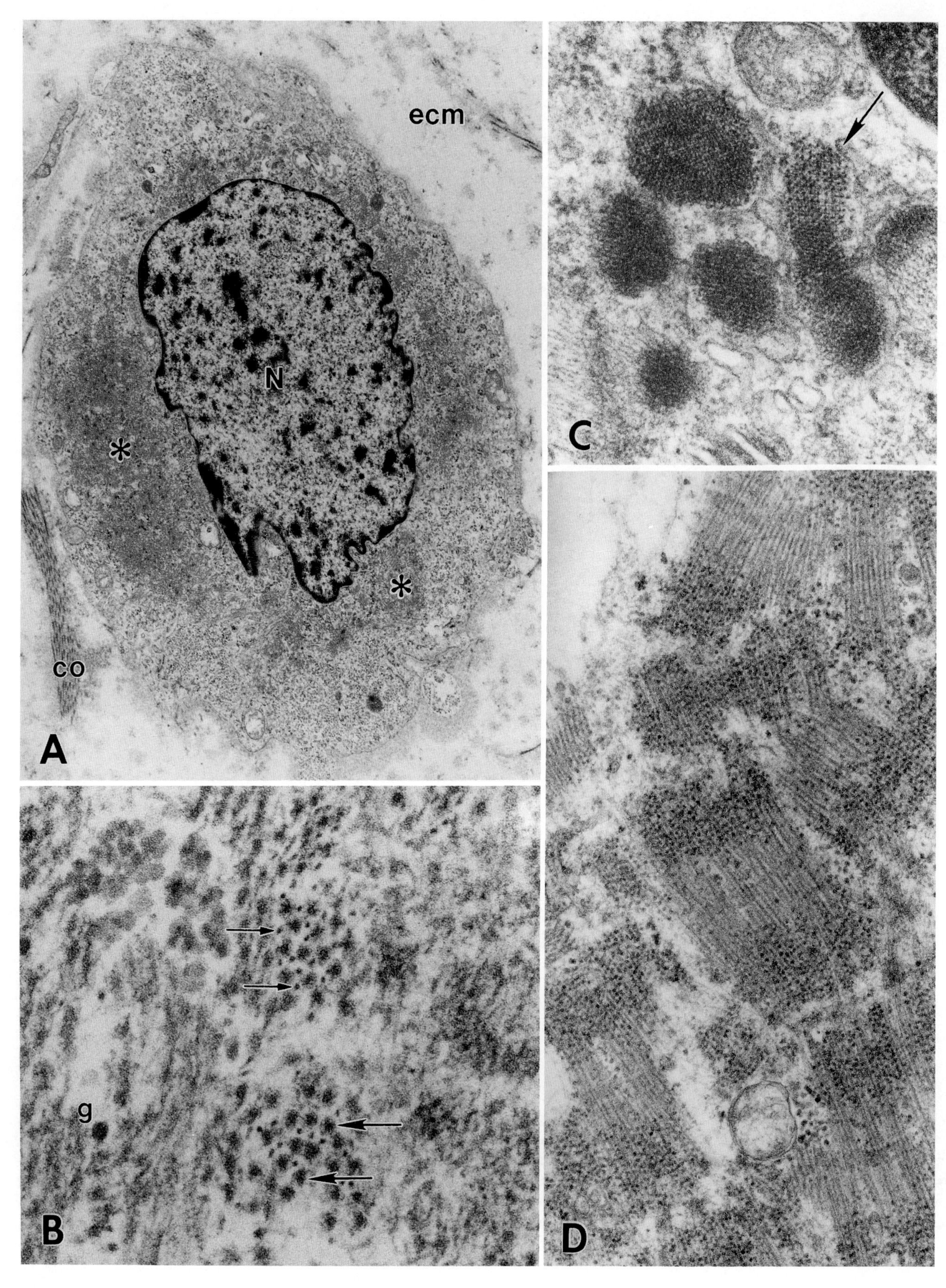
ecm
N
*
*
co
A
C
g
B
D

Plate 55. **A**: Tumor rhabdomyoblast with dense amorphous areas in cytoplasm (*). ×5400. **B**: Detail of amorphous area showing myosin (*large arrows*) and actin (*small arrows*) filaments without obvious geometrical arrangement. ×130,000. **A** and **B**: Same tumor as in Plate 53C. **C**: Cross-sectioned myofibrils, one showing quadratic arrangement of filaments (*arrow*). ×69,000. Same tumor as in Plate 54A. **D**: Ribosome–myosin complexes. Rhabdomyosarcoma, inferior vena cava. [Micrograph courtesy of Dr. Y. Kuwashima (Saitama, Japan). Reproduced from Kuwashima et al. (1992), with permission.] ccm, extracellular matrix; co, collagen; g, glycogen; N, nucleus.

23

Nemaline Rods: Leptomeric Fibrils

Nemaline rods (or **bodies**) and **leptomeric fibrils** are regarded as pathological alterations of the cytoskeleton in striated muscle cells and in certain tumors.

- *Nemaline rods* have been found in striated muscle cells, especially in myopathies and rhabdomyomas.
- *Leptomeric fibrils* have been seen in cardiac myocytes, rhabdomyomas, rhabdomyosarcomas, and the following epithelioid neoplasms: nevus of the iris, melanoma, hemangioendothelioma, and epithelioid sarcoma.

Nemaline rods are dense bodies, often rod-shaped and up to 1–2 μm across (Plate 56A). They are associated with sarcomeric myofilaments (Plate 56B) and have longitudinal and transverse periodicities. There is evidence that they are abnormal Z-disks (Fawcett, 1968); in fact, some Z-disks can show a nemaline-body-like transverse periodicity (Plate 56C).

Leptomeric fibrils are filamentous structures consisting of plates or bands of dense material—reminiscent of Z-disks—across which stretch fine filaments of actin size (Plate 57A and 57B). The dense bands have a regular spacing, usually of ~120 nm, but up to 300 nm has been recorded. Leptomeric fibrils have also been called **leptofibrils**, and aggregates of them have been referred to as **leptomeric complexes**.

Plate 56. **A** and **B**: Nemaline bodies in an abnormal striated muscle cell from scar tissue from an extraskeletal osteosarcoma, skin of neck. **A**, ×8000; **B**, ×60,900. In **A**, note the plate-like Z-disks in the adjacent cell; in **B**, note thick myosin (*large arrow*) and thin actin (*small arrow*) filaments and striations of nemaline body (*). **C**: Z-disk with transverse striations in a poorly defined myofibril. Same tumor as in Plate 54. ×69,000. N, nucleus; nb, nemaline body; Nu, nucleolus; Z, Z-disk.

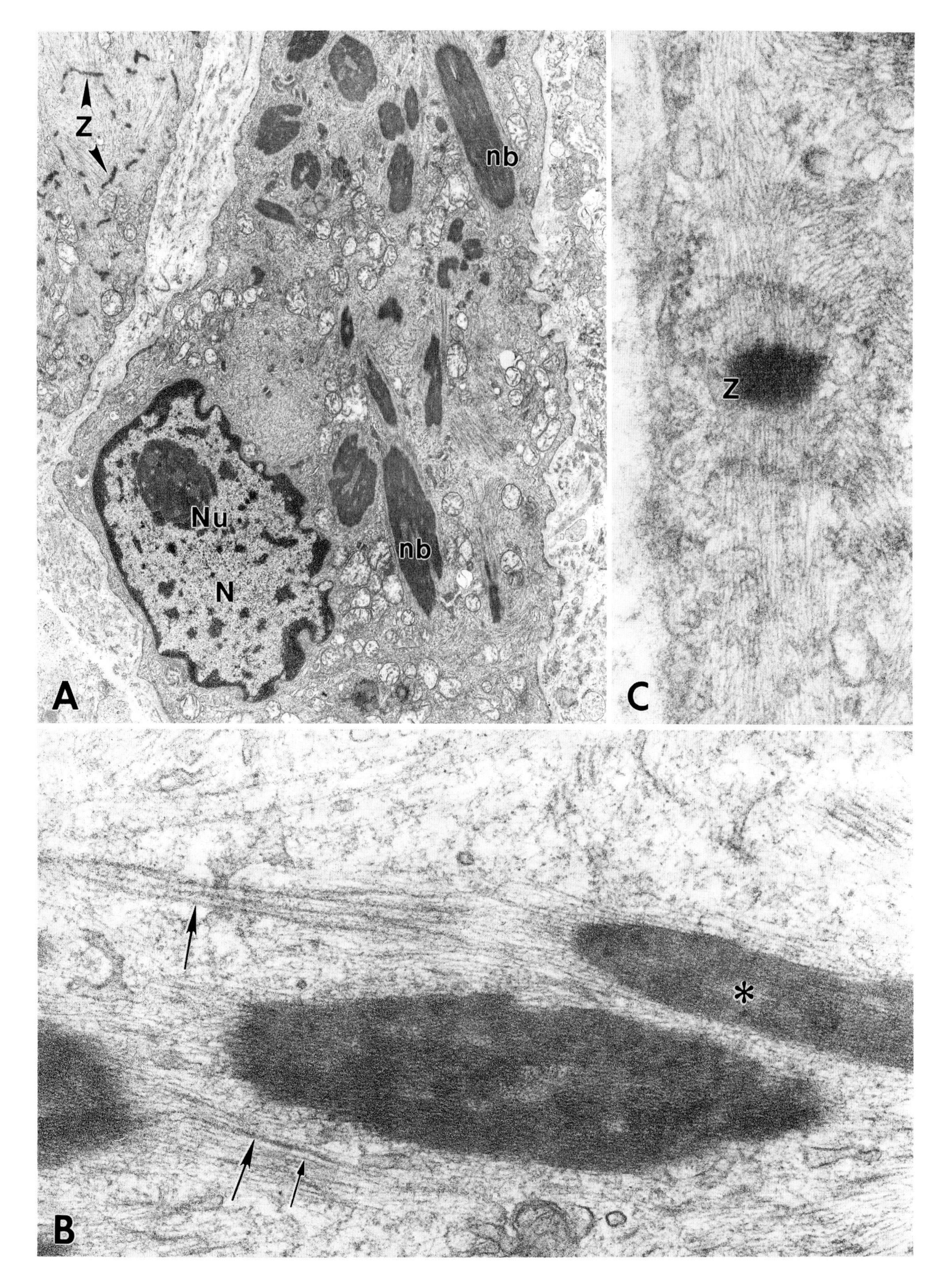
Z
nb
Nu
N
nb
A
Z
C
*
B

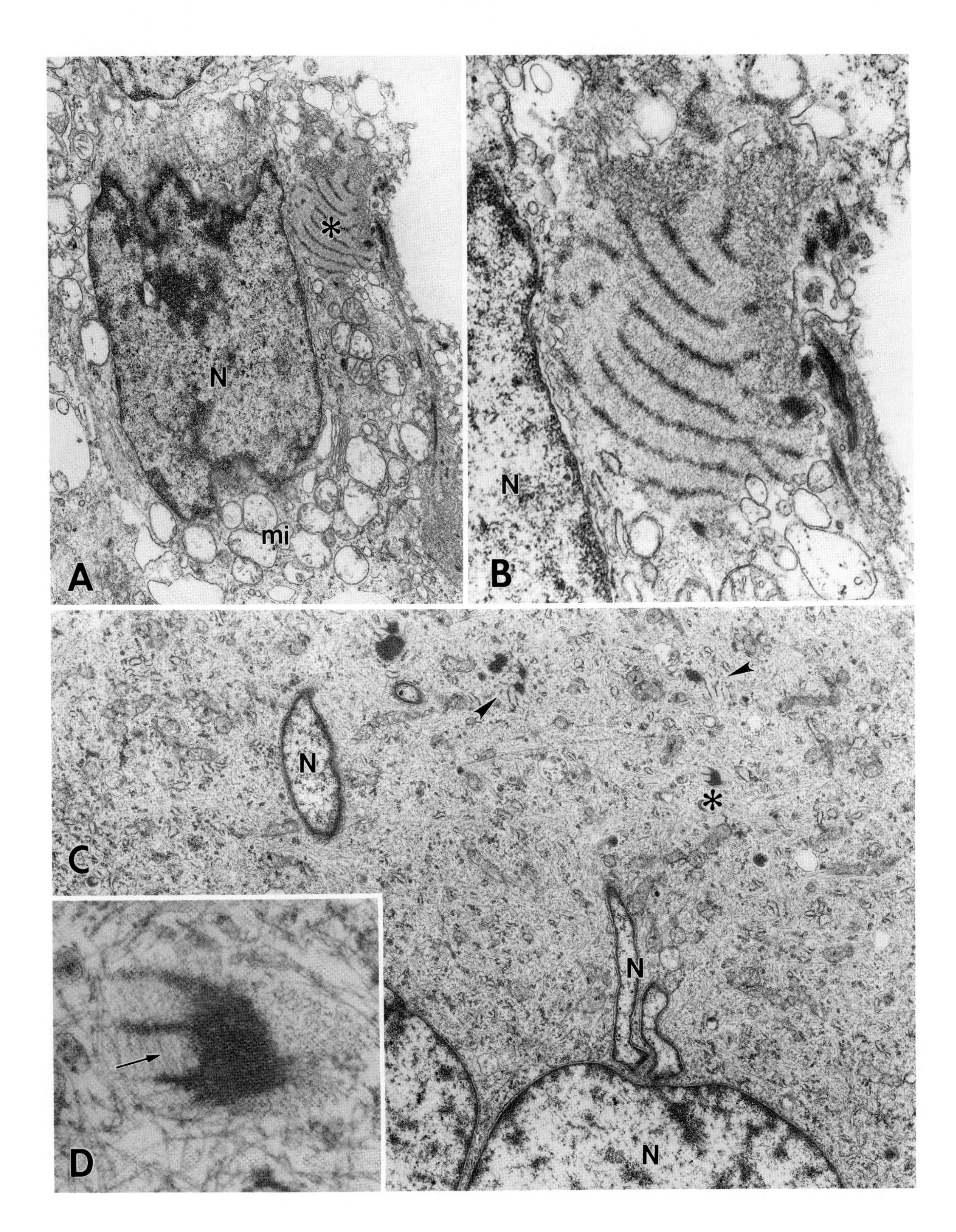

*
N
mi
A
N
B
N
N
*
C
N
N
D

Plate 57. **A**: Well-developed, rather broad leptomeric fibril (*) with a slightly larger than usual spacing of ~230 nm. Malignant peripheral neuroectodermal tumor with divergent immunophenotype, subcutis, calf. ×9900. **B**: Detail from **A**, ×29,300. **C**: Three small foci of leptomeric fibrils (*arrowheads* and *) in cytoplasm otherwise filled with intermediate filaments (by immunostaining these are a mixture of vimentin and desmin). Pleomorphic rhabdomyosarcoma, abdominal wall. ×8500. **D**: Detail of the asterisked fibril from **C** showing spacing of 150 nm and fine traversing filament (*arrow*). ×66,530. mi, mitochondria; N, nucleus.

24

Intermediate Filaments

Intermediate filaments are so-named because their diameter of ~10 nm is between that of actin filaments (5–7 nm) and that of myosin filaments (12–15 nm) and microtubules (25 nm). Except for the nuclear fibrous lamina (Chapter 2) which consists of intermediate filament proteins called **lamins** in *non*filamentous form, intermediate filaments are located in the cytoplasmic matrix between organelles where they interact with other components of the cytoskeleton.

There are several distinct classes of intermediate filaments which help to define different groups of normal cells:

Cytokeratins	Epithelial cells
Vimentin	Mesenchymal, lymphoid, and myeloid cells
Desmin	Muscle cells
Neurofilaments	Neurons
Glial filaments	Astrocytes and glial cells

Identifying these intermediate filaments by electron microscopy helps to define tumor cell differentiation and therefore has a role in diagnosis; however, with the partial exception of cytokeratin filaments, intermediate filaments are extremely difficult to distinguish from one another. Considerable reliance is placed on correlated light microscope immunostaining when immunoelectron microscopy is not available.

NONCYTOKERATIN INTERMEDIATE FILAMENTS

Most intermediate filaments—vimentin, desmin, neurofilaments, and glial filaments—have a similar appearance ultrastructurally. Their interpretation very largely depends on a preexisting knowledge of the cell type in which they are present, or on correlated light microscope immunostaining.

Intermediate filaments have a wide variety of appearances. They may be haphazardly arranged in small (i.e., normal) numbers lying between organelles (Plate 58A), but very often in tumors they accumulate in abnormal

Plate 58. **A**: Moderate numbers of wavy vimentin intermediate filaments between organelles. Note that the filament diameter is comparable to the thickness of the plasmalemma. Parotid cyst. ×60,700. **B**: Spheroidal aggregate of vimentin coexpressed with cytokeratin. Note selective orientation of some filaments (*arrowheads*). ×12,800. **C**: Same tumor as **B** in histologic section to show eosinophilic inclusions (*arrows*) due to intermediate filaments. ×410. **B** and **C**: Malignant rhabdoid tumor, bladder. [Reproduced from Harris et al. (1987), with permission from Blackwell Science, Ltd.] if, intermediate filaments; N, nucleus; pm, plasmalmma; rER, rough endoplasmic reticulum.

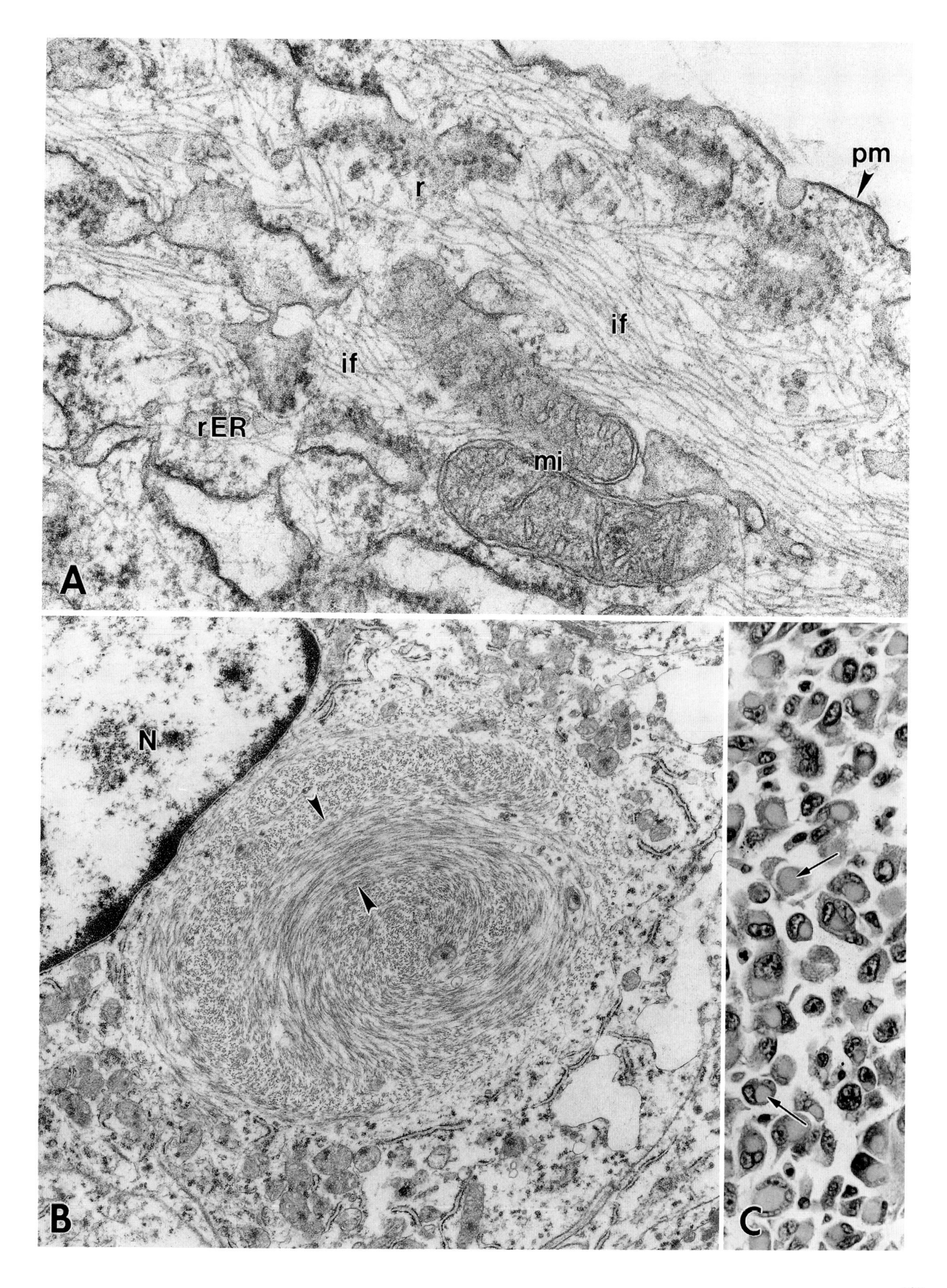
pm
r
if
if
rER
mi
A
N
B
C

numbers. In these circumstances, intermediate filaments may accumulate to form bands or spheroidal masses (Plate 58B). These masses may consist exclusively of filaments, or they may entrap organelles such as mitochondria, lipid droplets, or neuroendocrine granules. They are also the ultrastructural counterpart of some hyaline or eosinophilic inclusions seen by light microscopy (Plate 58C). At the extreme, almost the entire cytoplasm can be occupied with filaments (Plate 59A and 59B).

- Spheroidal masses of intermediate filaments can be found in almost all classes of tumor: carcinomas, neuroendocrine tumors, malignant melanomas, and sarcomas.
- They are required for the diagnosis of malignant rhabdoid tumors and are expected in Merkel cell tumors.
- Intermediate filaments are also a major component in the cortical Lewy body in neurons, Mallory bodies in liver, Rosenthal fibers in astrocytes, and neurofibrillary tangles (Lowe et al., 1988).

In intermediate filament aggregates, individual filaments are recognizable (Plates 58B and 59B), and although one can sometimes see groups which collectively orientate in a given direction (Plate 58B), they do not form discrete *bundles* like the cytokeratin tonofibrils of epithelial cells (see below). However, some filaments in tumors where immunostaining indicates vimentin but not cytokeratin form tonofibril-like bundles (Plate 59C and 59D). The fact that these **pseudotonofibrils** tend to appear in poorly preserved cells suggests that they may be artifacts resulting from the focal collapse of vimentin or other noncytokeratin filaments.

A note on terminology: The following terms have been used in the literature for intermediate filaments but should be avoided because they lead to confusion with actin filaments (Chapter 22): **thin filaments, microfilaments, type I microfilaments**, and **myofilaments**. The term **cytofilament** simply means "cell filament" and is too nonspecific to be useful.

CYTOKERATINS

Only the cytokeratins are identifiable in purely ultrastructural terms, and then only sometimes. Cytokeratins are recognized by their tendency to form linear bundles. Since cytokeratin filaments have traditionally been referred to as **tonofilaments**, a bundle of them is known as a **tonofibril**. Alternative but less suitable terms for cytokeratin include **pre-keratin** and **keratin**.

- Cytokeratins in the form of tonofibrils are found in many kinds of epithelial tumor, including neuroendocrine neoplasms, spindle cell carcinomas, monophasic and biphasic synovial sarcomas, epithelioid sarcomas, ameloblastomas, but especially squamous and basal cell carcinomas, and mesotheliomas.
- Tonofilaments are coexpressed with myofilaments in myoepithelial cell neoplasms and some squamous cell carcinomas.

Tonofibrils occur in two broad types. In nonsquamous epithelium and in mesothelium, tonofibrils consist of loosely organized and individually recognizable cytokeratin filaments, and the fibril is often of a moderate density (Plate 60A). In epidermis and tumors showing overt squamous and basal cell differentiation, tonofibrils typically show a high electron density (Plate 60B and 60C; see also Plate 66A and 66B). They also often present distinctively curving or splintering profiles (Plates 60C and 61) and frequently insert onto the plaques of desmosomes.

Sometimes, tonofibrils may be absent from the perinuclear cytoplasm (Plate 60B), and it is also rare for them to be found in subplasmalemmal locations; some subplasmalemmal smooth-muscle myofilaments with focal densities have a wavy arrangement mimicking tonofibrils. Dense squamous tonofibrils may also have a homogeneous texture in which individual filaments are not resolved, as if they have fused together (Plate 61C).

Cross sections of squamous tonofibrils may be angular and not be immediately apparent as a squamous feature (Plate 60C).

Plate 59. **A**: Cytoplasm of tumor cell almost fully occupied by filaments. Malignant fibrous histiocytoma, soft tissues, elbow. ×11,400. **B**: Detail of **A** showing typical size for intermediate filaments; these are probably vimentin on the basis of light microscope immunostaining in this case. ×23,400. **C**: Dense curving fibrils (*arrows*) reminiscent of true cytokeratin tonofibrils. Note the melanosome. Metastatic malignant melanoma, cervical node. ×95,300. **D**: Desmosome with attached structures resembling "tonofibril" tails (*arrows*). Meningioma, scalp. ×78,700. Tumors in **C** and **D** were poorly preserved (* indicates the exceptionally clear cytoplasmic matrix) and nonreactive for several commonly used anti-cytokeratin antibodies, but positive for vimentin. d, desmosome; m, melanosome.

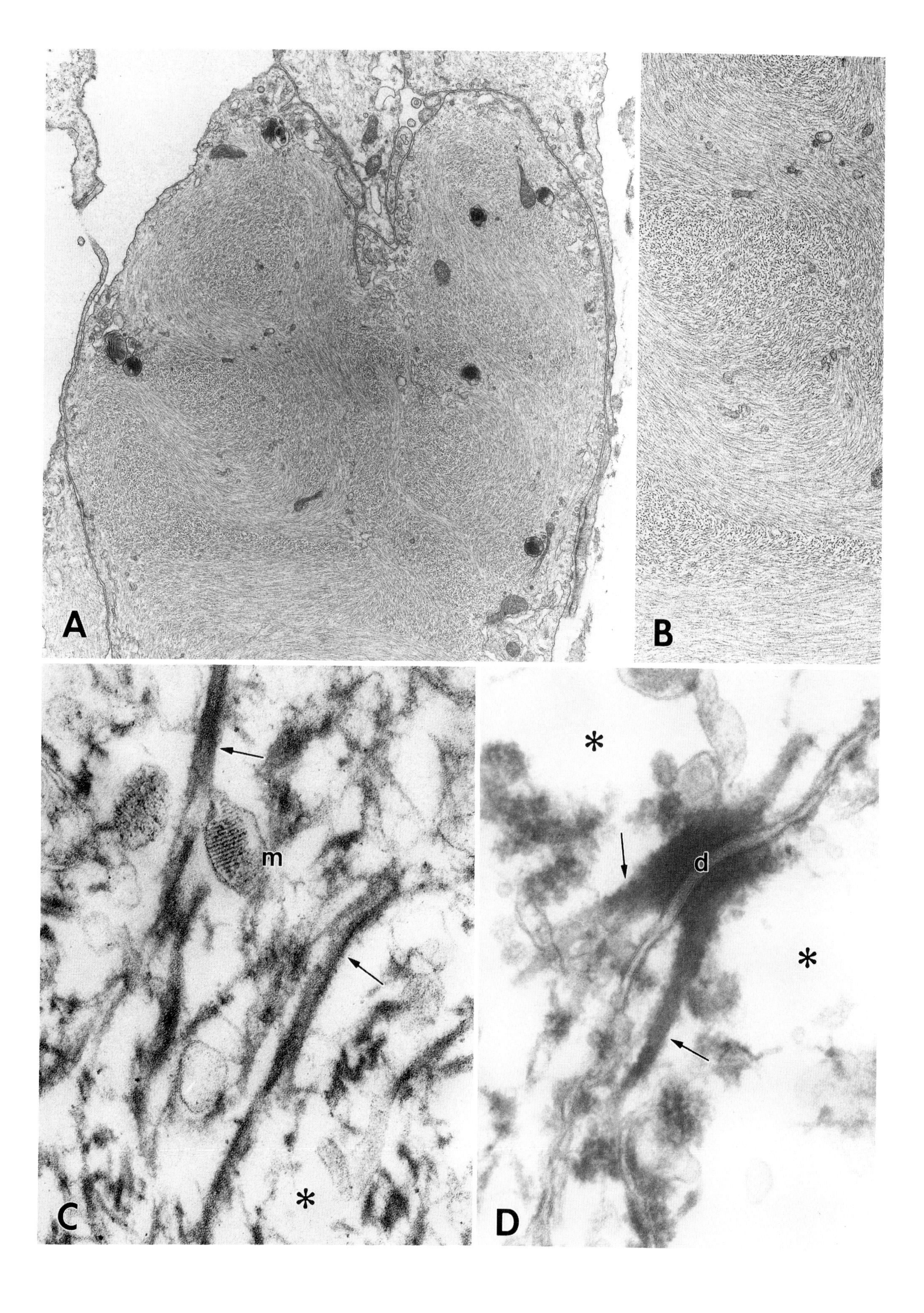

A
B
m
*
C
d
*
*
D

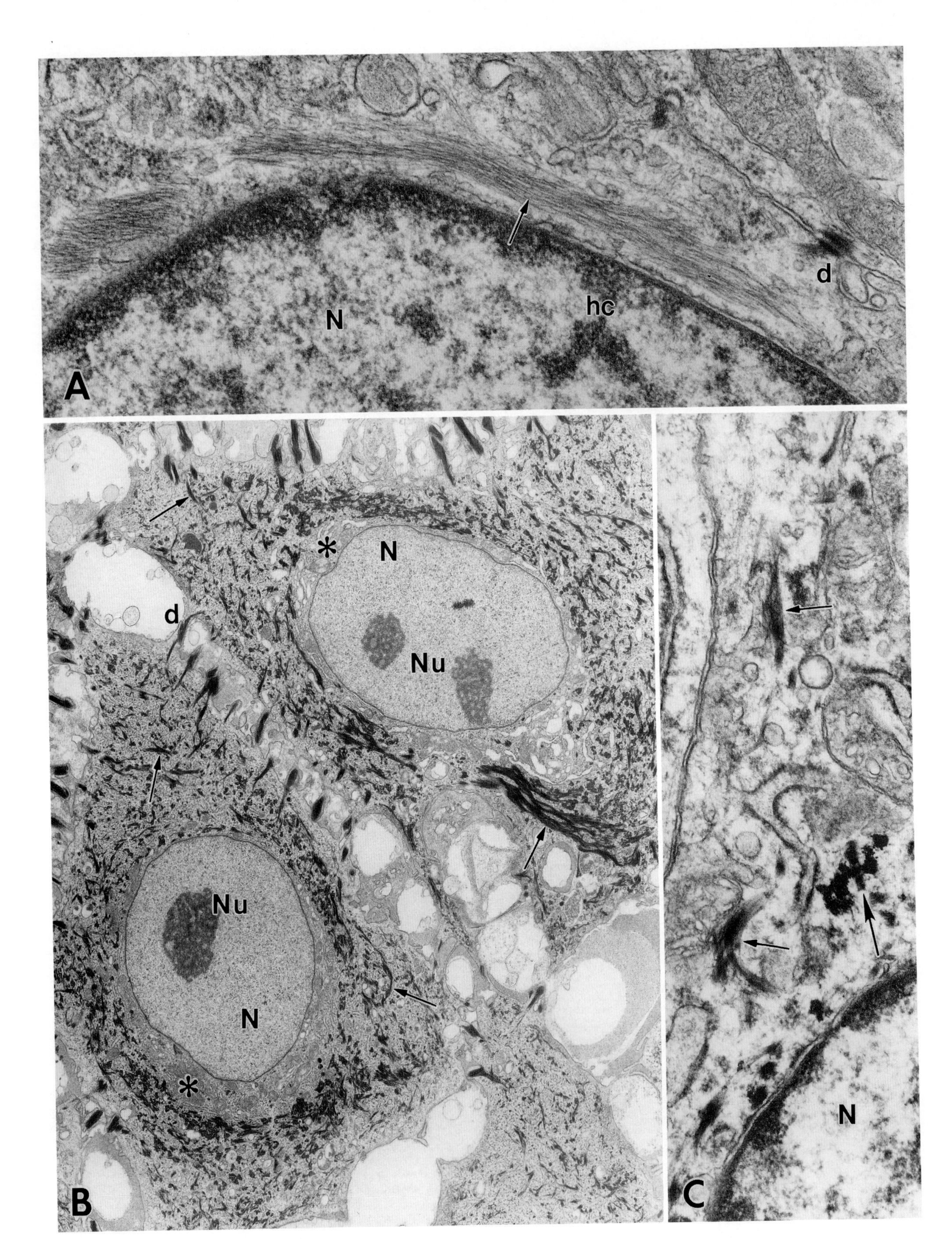
A
N
hc
d
B
N
Nu
d
*
Nu
N
*
C
N

Dense tonofibrils occurring in unambiguous adenocarcinoma indicate squamous metaplasia (Plate 61A), but some normal and neoplastic glandular epithelium can also show modestly developed tonofibrils. This may reflect either (a) artifactual collapse of "light-staining" tonofibrils or (b) subtle levels of squamous metaplasia.

Plate 60. **A**: Loosely organized tonofibril (*arrow*). Note desmosome. Also present is a nucleus containing coarsely granular heterochromatin. Epithelial mesothelioma, pleura. × 13,400. **B**: Dense squamous tonofibrils (*arrows*). The nuclei are extremely euchromatinic and are surrounded by a zone of cytoplasm (*) devoid of tonofibrils. Many desmosomes are present, but their structure is not obvious because of sectioning geometry. Anal epithelium. × 5300. **C**: Angular profile of cross-sectioned tonofibril (*large arrow*). Small arrows indicate small but typical dense tonofibrils. Spindle cell squamous carcinoma, skin of forehead. × 38,900. d, desmosome; hc, heterochromatin; N, nucleus; Nu, nucleolus.

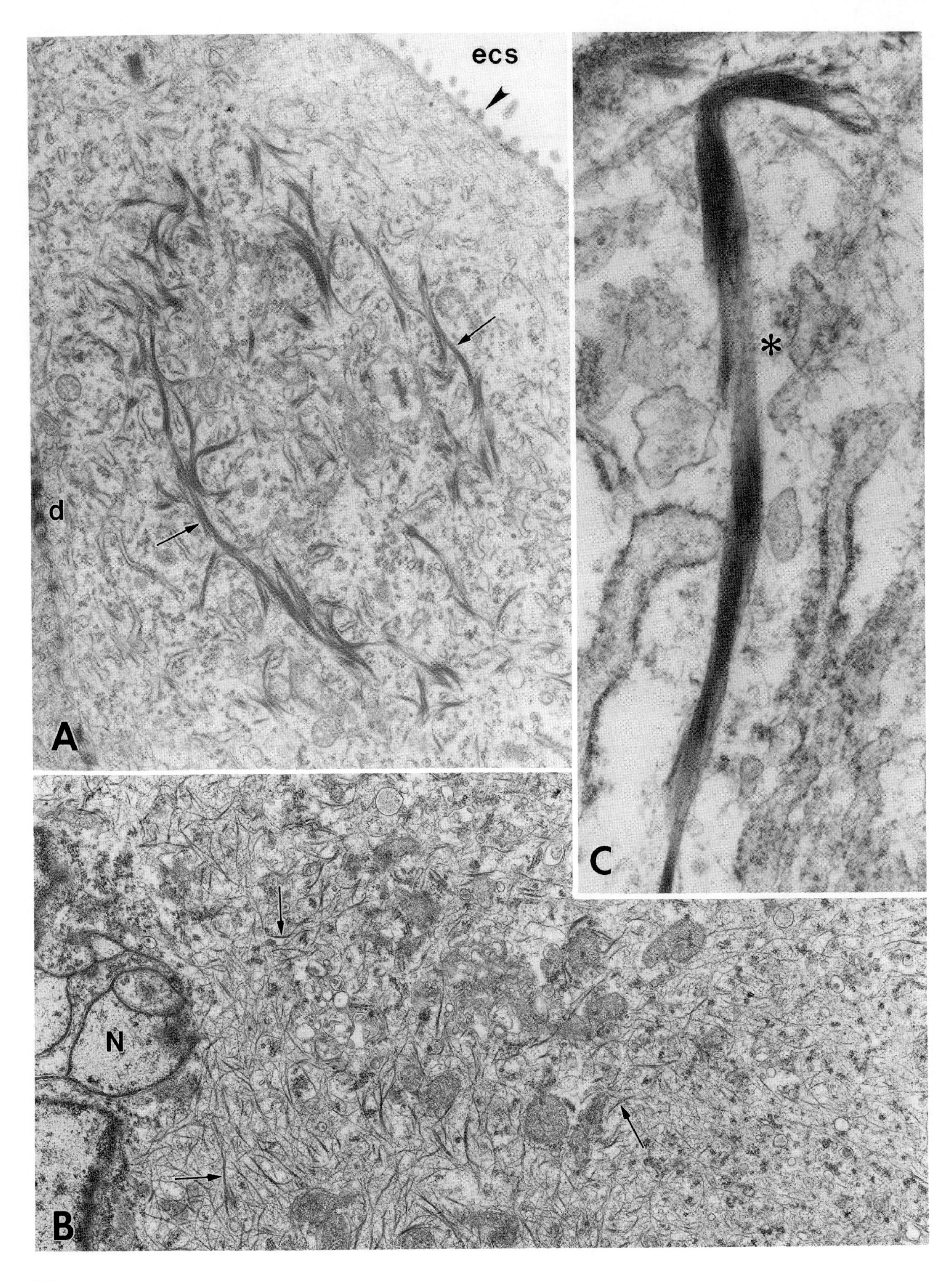
ecs
d
A
N
B
*
C

Plate 61. **A**: Typical dense squamous tonofibrils (*arrows*) in adenocarcinoma of endometrioid type, ovarian cyst. Note desmosomes and stubby microvilli (*arrowhead*). × 8500. **B**: Numerous very fine tonofibrils (*arrows*). Tumor cell showing myoepithelial differentiation in a carcinosarcoma, breast. × 13,300. **C**: Dense tonofibril showing homogeneous texture (*). Pseudoangiosarcomatous squamous cell carcinoma, breast. × 60,000. d, desmosome; ecs, extracellular space; N, nucleus.

25

Microtubules

The long 25-nm-diameter tubules found in the cytoplasmic matrix between organelles are referred to as **microtubules**. They are composed of subunits of tubulin protein and are the only major component of the cytoskeleton not generally referred to as filaments. Microtubules are only one of a number of structures having a tubular organization. Any structure with a tubular appearance and a diameter outside 15–30 nm is probably not a true tubulin microtubule; even some structures within this range do not have a proven tubulin composition and thus should preferably be referred to as **microtubular structures** (Chapters 3 and 11) rather than *microtubules*. Microtubules are found as free structures or organized into distinctive organelles: ciliary axonemes and basal bodies, the flagellar axoneme of the spermatozoon; centrioles; mitotic spindles; and the constriction between dividing cells known as the *midbody*.

FREE MICROTUBULES

In small or moderate numbers, free microtubules are almost ubiquitous in cells, where they function in cell movement and the intracellular translocation of organelles. They may, for example, be found near the surfaces of cells secreting collagen secretion granules (Plate 30C) and in the vicinity of the Golgi apparatus: here, they do not necessarily indicate neural/neuronal differentiation (see below). In larger numbers they align in the cell processes of neuronal and neuroendocrine cells and tumors (Plate 62). More rarely, they aggregate to form masses excluding other organelles (Pearl et al., 1981).

In large numbers, free microtubules are typical of: neurons (Plate 62) and neuronal tumors such as neuroblastomas and gastrointestinal autonomic nerve tumors; some Schwann cells; some carcinoids; certain central nervous system neoplasms; neuroendocrine tumors showing epithelial differentiation (e.g., neuroendocrine carcinoma).

MICROTUBULAR ORGANELLES

Mitotic spindles, centrioles, and the midbody of separating postmitotic cells (*Flemming body*) are nonspecific. Cilia, however, characterize luminal surfaces

Plate 62. **A–C**: Microtubules (*small arrows*) in neuronal processes (*large arrows*) from a differentiating neuroblastoma, axilla. **A**, transverse (×66,530); **B**, oblique (×69,600); and **C**, longitudinal (×66,530) section. In **C**, the small vacuoles/vesicles seem to have too tough a membrane to be synaptic vesicles and may represent cisternae of smooth endoplasmic reticulum. p, process; pm, plasmalemma; v, vesicle.

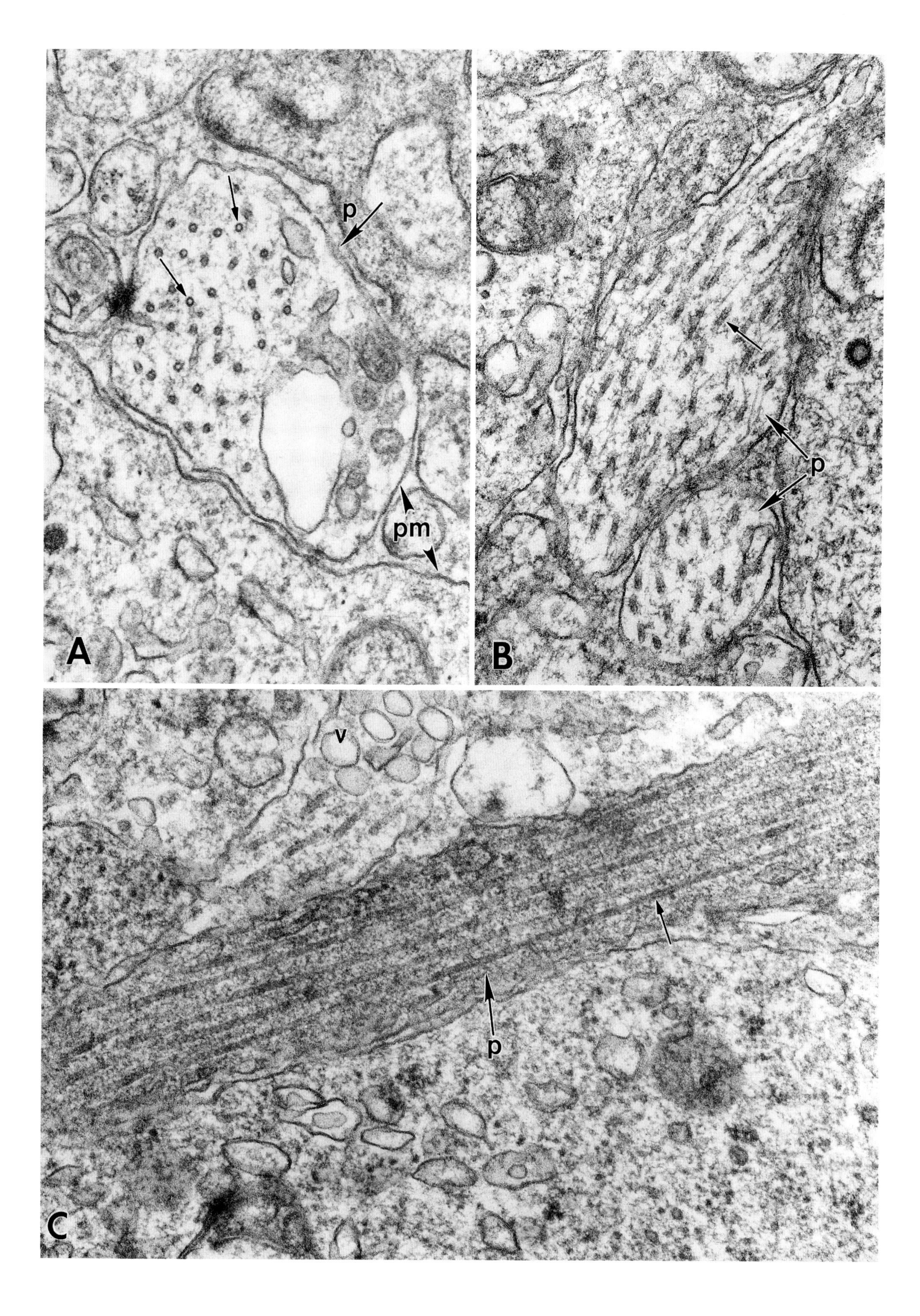
p
pm
A
p
B
v
p
C

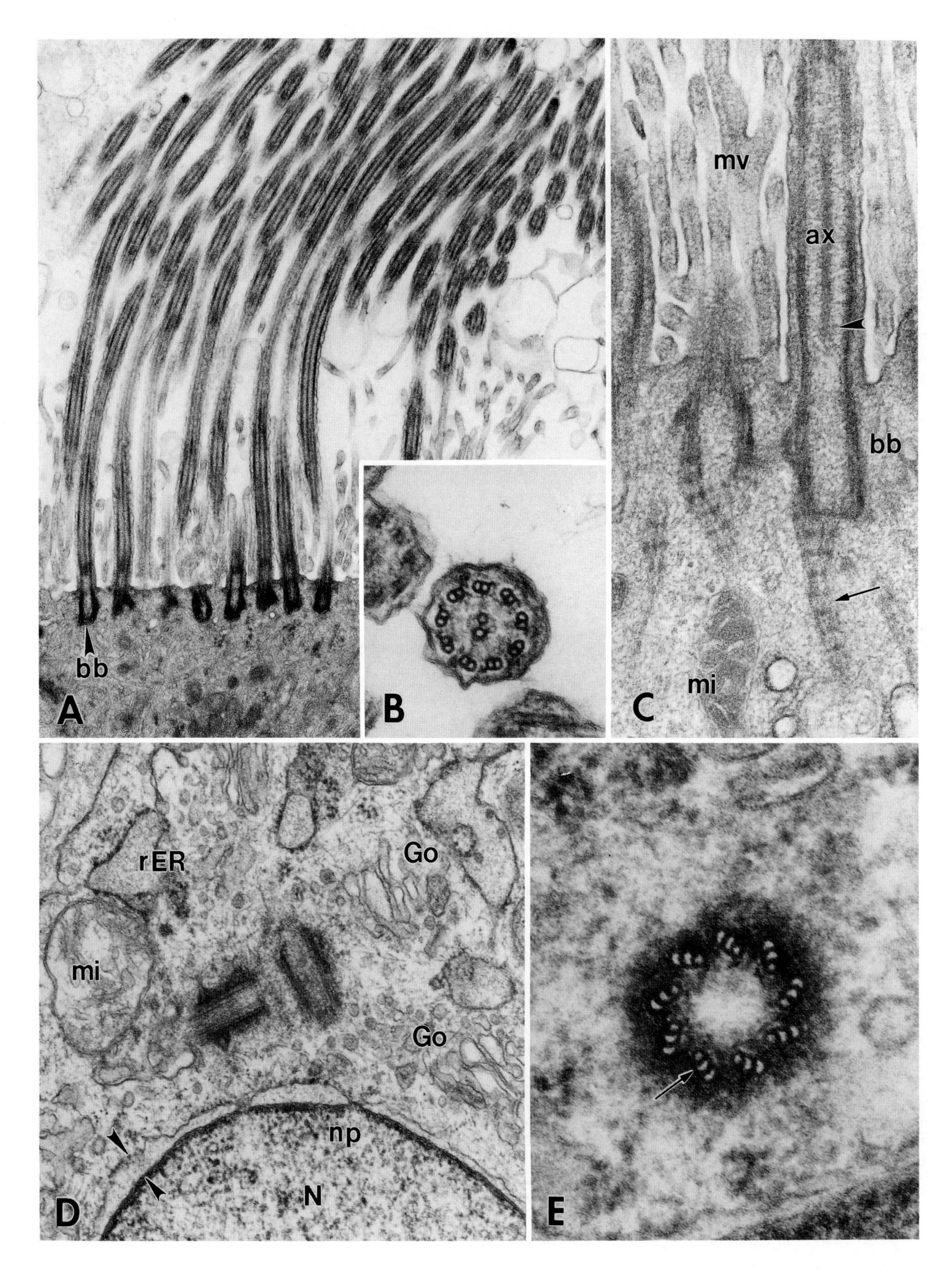
bb
A
B
mv
ax
bb
mi
C
rER
Go
mi
Go
np
N
D
E

from the upper respiratory and reproductive tracts, as well as certain other cells such as in ependymomas (these tumors may also have multiple basal bodies).

The **ciliary axoneme** (Plate 63A and 63B) consists of nine peripheral microtubule doublets and two central separate microtubules. The organization is best appeciated in transverse section. In long section only a proportion of the axonemal microtubules are seen, and often they go in and out of the plane of section. Although they are not significant in tumor diagnosis, ciliary axoneme abnormalities have been described in which microtubule linkages may be abnormal or absent, or where microtubules may be deficient or supernumerary and are considered to be responsible for ciliary immotility (Lurie et al., 1992; Mierau et al., 1992).

The basal body (Plate 63C) has nine peripheral microtubule triplets and no central microtubules; it may have well-developed cross-striated rootlets anchoring the basal body into the cytoskeleton (Plate 63C).

Centrioles (Plate 63D and 63E) have a similar organization to basal bodies but are found in pairs close to the nucleus and Golgi apparatus. Only in favorable sections does one see the typical mutually orthogonal arrangement. Centrioles may have associated cross-striated fibrils, presumably acting as anchoring devices. Frequently, even in normally nonciliated cells and often in neoplastic cells, one of the centrioles extends to produce what is referred to as an **oligocilium** (Gonzalez et al., 1985).

Aggregates of microtubules and tubulin crystals can be encountered in the cytoplasm (El-Labban et al., 1988). They lack defined functions. Microtubules survive paraffin embedding.

Plate 63. **A–C**: Cilia from normal nasal mucosa. **A**: Distribution and coexpression with microvilli. ×13,280. **B**: Cross section of ciliary axoneme. ×93,200. **C**: Ciliary basal body (note discontinuity of central microtubules at *arrowhead*) and cross-striated rootlet (*arrow*). ×60,800. **D**: Typical arrangement of centrioles near Golgi and nucleus. A nuclear pore and the nuclear envelope are clearly delineated. Malignant peripheral neuroectodermal tumor with divergent immunophenotype, subcutis, calf. ×39,600. **E**: Cross section of centriole showing nine microtubule triplets (*arrow*). Fibrosarcoma, biceps muscle, ×140,000. bb, basal body; ax, axoneme; Go, Golgi apparatus; mi, mitochondrion; mv, microvilli; N, nucleus; np, nuclear pore; rER, rough endoplasmic reticulum.

26

Intercellular and Cell-to-Matrix Junctions

Intercellular junctions (junctions *between* cells) provide physical cohesion in tissues, facilitate biochemical communication, and create watertight domains. **Cell-to-matrix junctions** also contribute to tissue integrity by promoting adhesion between cells and matrix constituents, as well as being involved in the dynamic processes of tissue remodeling (e.g., the fibronexus; see Chapter 27).

Junctions may be solitary (desmosomes, tight junctions, gap junctions, subplasmalemmal linear densities) or composite (junctional complex, intercalated disk). Some are of considerable value in tumor diagnosis in contributing information on cellular differentiation, whereas others are nonspecific.

The most diagnostically important junctions are desmosomes, tight junctions, junctional complexes, subplasmalemmal linear densities, and fibronexus junctions (Chapter 27).

DESMOSOMES

The **desmosome** (Plates 59D and 60A) is one of the most important markers of epithelial differentiation. It is found in nearly all normal and neoplastic epithelial tissues; exceptions include poorly differentiated tumors and, for example, adrenal cortex.

Desmosomes are important in the diagnosis of squamous, glandular, and neuroendocrine carcinomas, as well as those tumors not considered primarily epithelial but showing a measure of epithelial differentiation or architecture—for example, meningiomas, mesotheliomas, ameloblastomas, chordomas, granulosa cell tumors, and rhabdomyomas. Rare desmosomes have been seen in malignant melanomas and peripheral primitive neuroectodermal tumors.

Ultrastructure of Desmosomes

Desmosomes have a number of distinctive features. Plates 19 and 60 give a perspective on their distribution amongst cells and their size. They have the following features: (1) uniform width of about 20–30 nm between the apposed plasma membranes; (2) an **intermediate linear density** or **line** running down the middle of the intercellular space; (3) sharply delineated **plaques** of uniform thickness (about 20–40 nm), consisting of amorphous dense material located immediately beneath the membranes (Plate 64A and 64B); (4) cytokeratin filaments (tonofilaments) attached to these plaques (Plate 64A): these are **tonofilament** or **tonofibril tails** (Plates 59D, 60A, and 64A).

Several other points are worth making in connection with desmosomes: (1) The intercellular space within the domain of the plaques is often wider than that in the immediate vicinity of the junction (Plate 64A). (2) Sometimes the plaque and the associated cytokeratin filaments are found only on one of the two apposed membranes; these are **half-desmosomes**. (3) Desmosomes can be found within the cytoplasm without any apparent contact with the cell surface and are referred to as **intracellular desmosomes**. Some of these are probably intended for internal digestion, but some *apparently* intracellular desmosomes may be artifacts of sectioning geometry. (4) Although most desmosomes are associated with tonofilaments which are cytokeratin in nature, some identical junctions are associated with vimentin—for example, in human lens epithelium—and can be called **vimentin desmosomes**. (5) Desmosomes are imaged in thin sections as linear structures but are really plate-like and rounded; a hint of this can be obtained in tangential sections. Most desmosomes have linear profiles in the range of 100–500 nm but they can reach about 1 μm, and a few rare examples exceed 1 μm. There is a suggestion that the desmosomes in epithelial malignant mesothelioma are bigger than those in adenocarcinoma (Burns et al., 1988). (6) Desmosomes may form rows connected by tonofilament bundles; they may be formed between the folded plasma membranes of the same cell (these are **autodesmosomes**). Desmosomes can survive paraffin embedding, although some of their characteristic structure may be lost.

Simplified or Primitive Desmosomes

Some junctions possess ultrastructural similarities to desmosomes but lack the full complement of features. Some contain linearly arranged material in the intercellular space and subplasmalemmal dense material; this may be loosely organized and not sharply defined as in true desmosomal plaques, and tonofibril tails may be absent (Plate 64C). It is difficult to know whether these are desmosomes or not. If they occur in a tumor, where on clinical, histological, and immunocytochemical grounds there is strong evidence of epithelial differentiation, there will be a tendency to want to interpret them as **primitive** or **simplified desmosomes**. Under these circumstances, the immunoelectron microscopic demonstration of desmosome plaque proteins and cytokeratin would help to provide a more convincing interpretation (Moll et al., 1986).

Primitive Junctions

Among the commonest of junctions are those which have no intermediate line and no associated filaments. They are characterized only by a close approach of membranes and subplasmalemmal material, usually amorphous and not organized as sharply defined plaques (Plate 64D and 64E). These junctions can be referred to as **simple** or **primitive junctions**.

Primitive junctions are present in almost all classes of tumor: carcinomas, neuroendocrine tumors, melanocytic lesions, neuronal tumors, sarcomas, and even some lymphomas.

Although primitive junctions in large numbers would be unexpected in a lymphoma, apart from this instance they are not significant diagnostically. The point of mentioning them is that they are often described in the literature as desmosomes, with the inevitable implication of epithelial differentiation. They are certainly not markers of epithelial differentiation, and they should not be referred to as desmosomes. In fact, the extent to which these junctions resemble true desmosomes in biochemical terms is almost completely unknown. For this reason, the other names frequently used for them—**desmosome-like junction** and **desmosomal junction**—should be avoided. **Intermediate junction** and **adherent junction** are also unsuitable for this type of junction.

JUNCTIONAL COMPLEX

Glandular epithelial cells are joined at their apical regions near the lumen by adhesive devices called **junctional complexes**. They are equivalent to the **terminal bars** of light microscopy.

In combination with a lumen and microvilli, the junctional complex is one of the most important ultrastructural markers for glandular epithelial differentiation in normal and neoplastic tissues. It is found in adenomas and adenocarcinomas, adenoid cystic carcinomas, mesotheliomas, biphasic synovial sarcomas, and Wilms' tumor.

Ultrastructure of the Junctional Complex

The junctional complex in its fully developed form consists of three distinct junctions (Plate 65A and 65B): a **tight junction** closest to the lumen, which creates a watertight seal; a **desmosome** furthest away, which promotes cellular cohesion, and between these two an adhesive device called the **intermediate junction** (Plate 65B).

In normal glandular epithelium and well-differentiated glandular epithelial tumors—that is, where there are well-defined lumina—the tight junctions

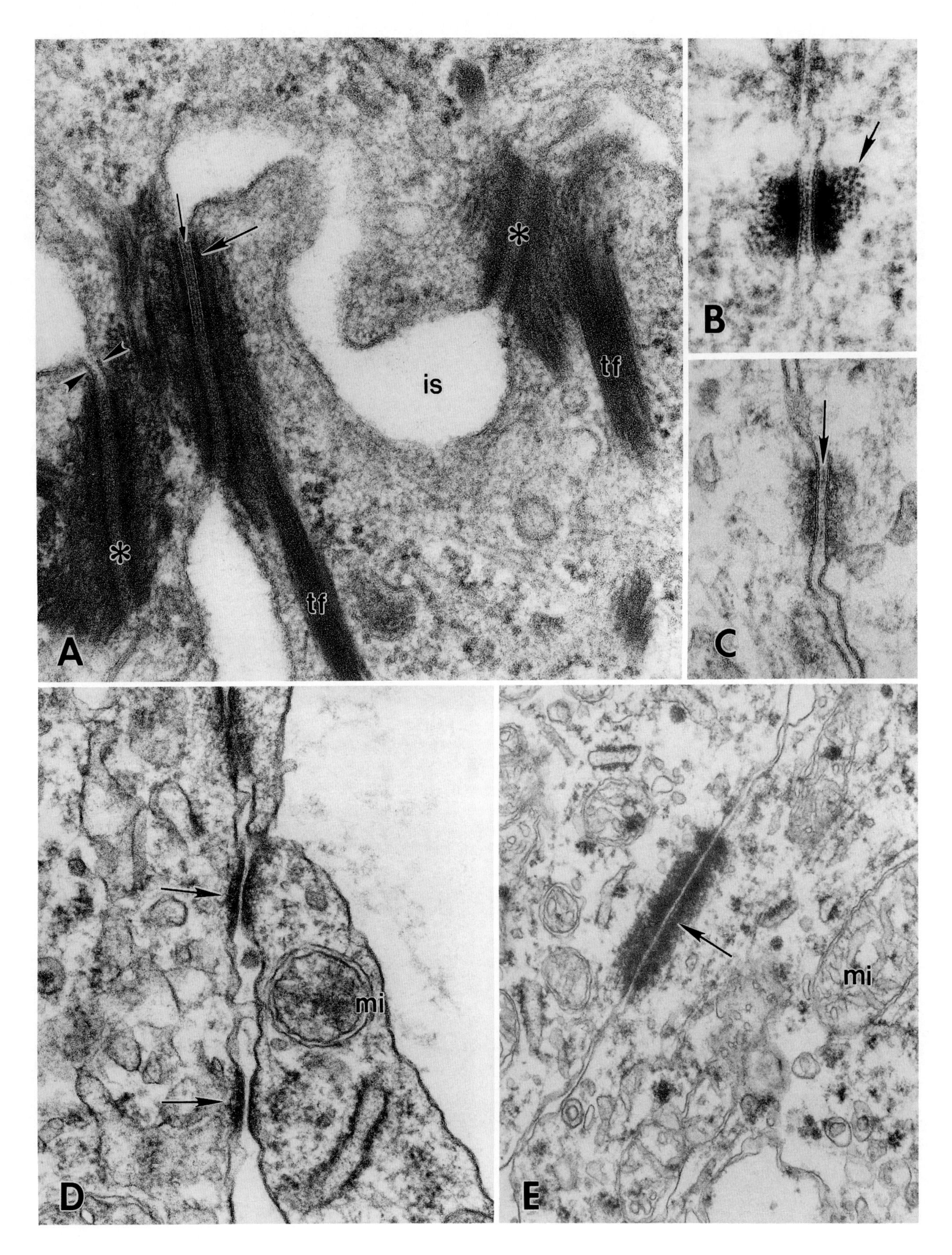
A
B
C
D
E
is
tf
tf
mi
mi

and intermediate junctions form belts (**zonules**) around the lumina. The desmosomes, however, are always localized, individual, and spot-like and are sometimes described as **macular**. In addition, the tight junctions and the desmosomes in junctional complexes have more or less the same structure as those found elsewhere (for example, in endothelium and nonglandular epithelium, respectively). Because the **intermediate junction** is defined as the middle adhesive device of the junctional complex, strictly speaking it does not have an unambiguous counterpart as an *isolated* junction in tumors. It differs from the desmosome (the other adherent junction in the junctional complex) by lacking material in the intercellular space and by the fact that the dense subplasmalemmal material is a mat of tangled nonmuscle actin filaments forming part of the terminal web rather than plaques.

Variations in junctional complex organization can be found, particularly in neoplastic tissues. On a rare occasion, tight junctions may be found on both sides of the desmosome, but more commonly the desmosome may be missing; the junctional complex then consists of a tight junction combined with an intermediate junction (Plates 65A and 71B). These junctions should still be regarded as junctional complexes (although incomplete ones) because of their physical association with lumina. It is worth pointing out that sometimes intermediate junctions assume particularly long profiles, reflecting their belt-like organization (Plate 65A).

Tight Junctions

In tight junctions, membrane apposition is so close that it completely excludes intercellular space; tight junctions therefore create watertight domains and prevent fluid loss to stroma from luminal compartments.

Tight junctions are found between endothelial cells (Plate 65C) and between Sertoli cells and are the most adluminal component of the junctional complex; they are diagnostically important, therefore, in vasoformative tumors, Sertoli cell tumors, and adenomas and adenocarcinomas.

Structure of Tight Junctions. Tight junctions are not too easy to demonstrate by purely structural criteria because high magnification and good membrane definition are required (Plate 65C). In ultrathin sections, tight junctions are characterized by linear expanses of apposed membrane with no intercellular space (Plate 65B) or points of contact with a cross-like appearance (Plate 65C). Often, however, tight junctions are interpreted on low-power micrographs where the above details are not resolved, with the interpretation being based on a knowledge of the cell concerned; for example, most junctions seen in endothelial cells are probably tight junctions.

Gap and Septate Junctions

Gap junctions and **septate junctions** mediate biochemical communication. Gap junctions, for example, appear in large numbers between myometrial smooth-muscle cells at the onset of contraction, thereby permitting the entire population of cells to act as a single functional unit (Garfield and Hayashi, 1981). A gap junction may also be referred to as a **nexus** or **nexus junction**, but these are somewhat old-fashioned terms.

- Gap junctions are nonspecific, occurring in both epithelial and mesenchymal cells; they have not been recorded in lymphoid cells.
- Septate junctions are well known in invertebrate tissues but are less well documented in human tissues; examples include a deep fibrous histiocytoma, ameloblastoma, normal urothelium, and abnormal erythroblasts.

Gap Junction Structure. Gap junctions have closely apposed membranes without associated filaments or dense material (Plate 65D–F). The 2-nm space is not always easy to see (Plate 65F) and may require an electron-dense tracer such as lanthanum nitrate to demonstrate it. The difficulty in seeing the space can be due to excessive membrane staining or sectioning

Plate 64. **A**: Well-developed desmosomes. Note intermediate line (*small arrow*), plaques (*large arrow*), and tonofilament bundles. Observe also restriction of intercellular space at the end of the plaques (*arrowheads*). Two desmosomes (*) lie at a slight angle within the section so that the plaques and intercellular spaces are not sharply imaged. Keratinizing squamous cell carcinoma, skin of neck. ×63,000. **B**: Small but well-formed desmosome where the associated tonofilaments are cross-sectioned (*arrow*). Poorly differentiated (microglandular) adenocarcinoma, eyelid. ×120,500. **C**: Junction considered as a primitive desmosome. Note material suggestive of intermediate line (*arrow*). Myoepithelial cell tumor, breast. ×95,300 [Reproduced from Pitt et al. (1995), with permission from Blackwell Science, Ltd.] **D**: Simple junctions (*arrows*) between a cell process and a cell body. Aggressive angiomyxoma, buttock. ×61,100. **E**: Primitive junction. In spite of the fact that this is a junction of substantial size, the subplasmalemmal dense material (*arrow*) is amorphously organized (not plaque-like) and the features are not those of a desmosome. Gastrointestinal autonomic nerve tumor, omentum. ×39,900. is, intercellular space; mi, mitochondrion; tf, tonofibril.

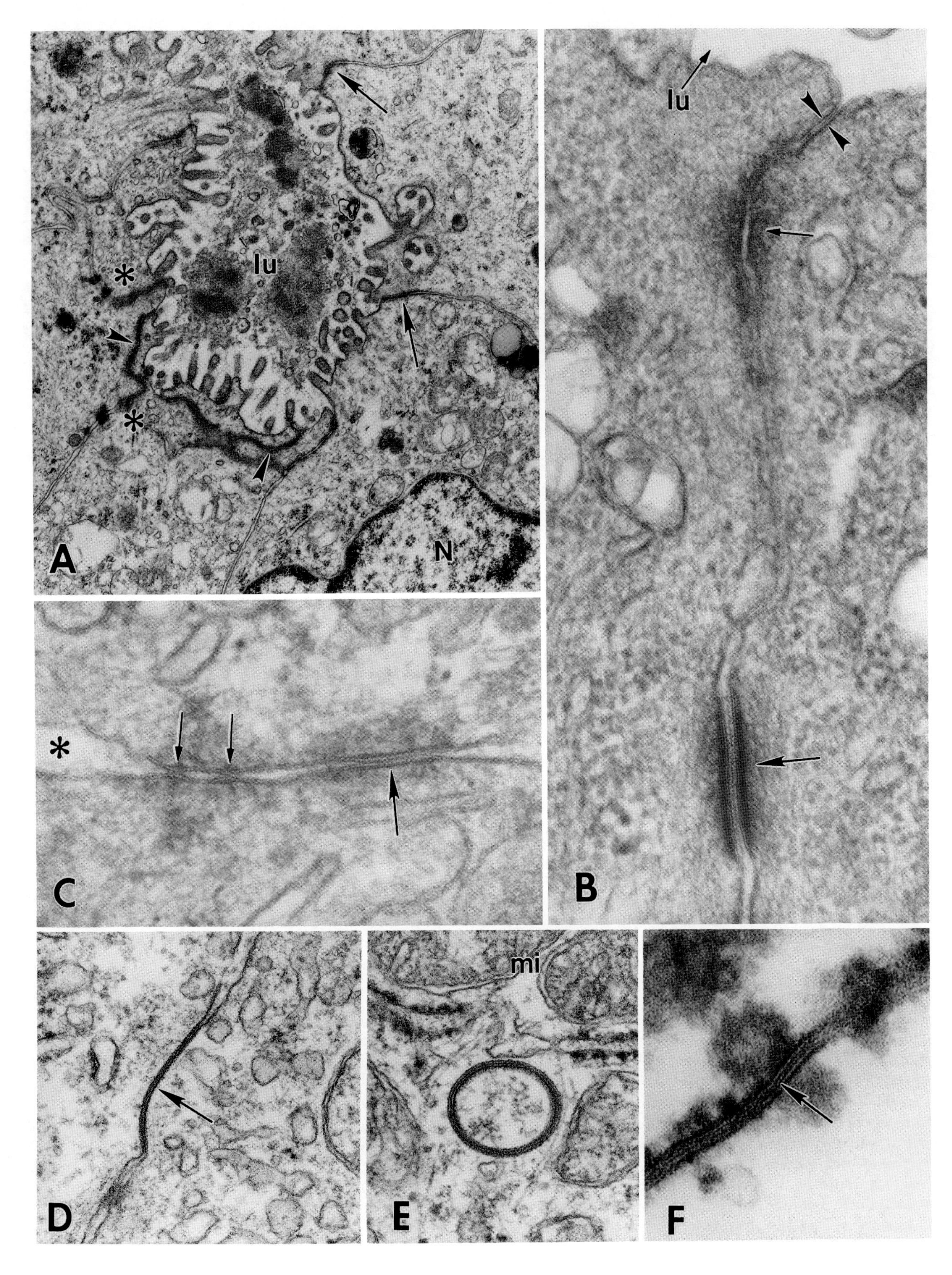
A
lu
N
B
lu
C
D
E
mi
F

geometry, or the space may be lost through artifactual collapse of the two apposed membranes. Gap junctions, however, often display a characteristic gently curving profile (Plate 65D). One abnormal variant is completely intracytoplasmic and spherical (Plate 65E).

Junctional Complex Terminology

The terminology for junctions within the junctional complex is complicated in the literature by the fact that Latin names for these junctions also exist. These are given here for completeness and because they are encountered in the literature, but they are best avoided because they provide an extra source of terminological confusion and have cumbersome plural forms:

Tight junction	Zonula occludens (plural: zonulae occludentes)
Intermediate junction	Zonula adherens (plural: zonulae adherentes)
Desmosome	Macula adherens (plural: maculae adherentes)

The **intercalated disk** of striated muscle tissue is not of major diagnostic importance because it is readily observed by light microscopy; see McNutt (1970) for ultrastructural details.

HEMIDESMOSOMES; SUBPLASMALEMMAL LINEAR DENSITIES: PAIRED SUBPLASMALEMMAL LINEAR DENSITIES

Hemidesmosomes, subplasmalemmal linear densities, and **paired subplasmalemmal linear densities** are specializations of the cell surface which have roles in cell-to-matrix adhesion or interaction. **Lamina** and the **fibronexus junction** are further examples of cell-to-matrix specializations and are discussed in Chapter 27.

- Hemidesmosomes are found on certain types of epithelial and myoepithelial cells abutting stroma (they are best known in epidermis, mammary gland, prostate, and endometrium and can be expected in tumors occurring at these sites).
- Subplasmalemmal linear densities and paired subplasmalemmal linear densities are markers of mesenchymal cells (including macrophages) and their derived tumors.

Ultrastructure of Hemidesmosomes

Hemidesmosomes resemble desmosomes in possessing dense material attached to the undersurface of the plasma membrane, with which tonofilaments are associated (Plate 66A–C). However, in contrast to the plaques of desmosomes, the dense material is often rather disorganized (Plate 66C): to a certain extent, this is due to the fact that hemidesmosomes are often positioned on rounded or narrow cellular processes which jut out into the matrix. Although there is no intermediate line, there is a thin and inconspicuous electron-dense plate just a few nanometers away from the external surface of the plasma membrane (Plate 66C and 66D): this is the **subbasal dense plate.**

Hemidesmosomes, like all other areas of the basal surface, have an overlying basal lamina (Plate 66A–C) which often shows enhanced electron density in the immediate vicinity of the hemidesmosome (Plate 66C). The term *half-desmosome* is occasionally used in the literature for hemidesmosome but should be avoided.

Ultrastructure of Subplasmalemmal Linear Densities and Paired Subplasmalemmal Linear Densities

The **subplasmalemmal linear density** (SLD) consists of a localized area of dense material attached to the undersurface of the plasmalemma and a correspondingly

Plate 65. **A**: Intercellular lumen (containing cell debris) exhibiting several junctional complexes. Some are complete (with desmosomes) (*); others are incomplete without desmosomes (*arrows*). Note extended intermediate junctions (*arrowheads*). Biphasic synovial sarcoma, soft tissues of chest wall. ×7900. **B**: Detailed structure of junctional complex, showing tight junction (*arrowheads*), intermediate junction (*small arrow*), and desmosome (*large arrow*). The tight junction is sectioned along its length. Normal breast. ×95,200. **C**: A row of three tight junctions. Two (*small arrows*) are sectioned, giving a cross-like appearance; the other is sectioned along its length (*large arrow*). The space (*) is a lumen. Endothelial cell in a vessel from a malignant peripheral nerve sheath tumor, axillary soft tissues. ×141,000. **D**: Typical profile of gap junction (*arrow*). The intercellular space is not resolved at this magnification. The junction is about 0.5 μm across. Tumor diagnosed by light microscopy as neuroendocrine carcinoma (but lacking neuroendocrine granules), retroperitoneum. ×52,000. **E**: Spherical gap junction. Cutaneous leiomyosarcoma, pinna, ×69,000. **F**: Detail of gap junction showing intercellular 2-nm space (*arrow*). Meningioma, scalp. ×170,000. lu, lumen; mi, mitochondria; N, nucleus.

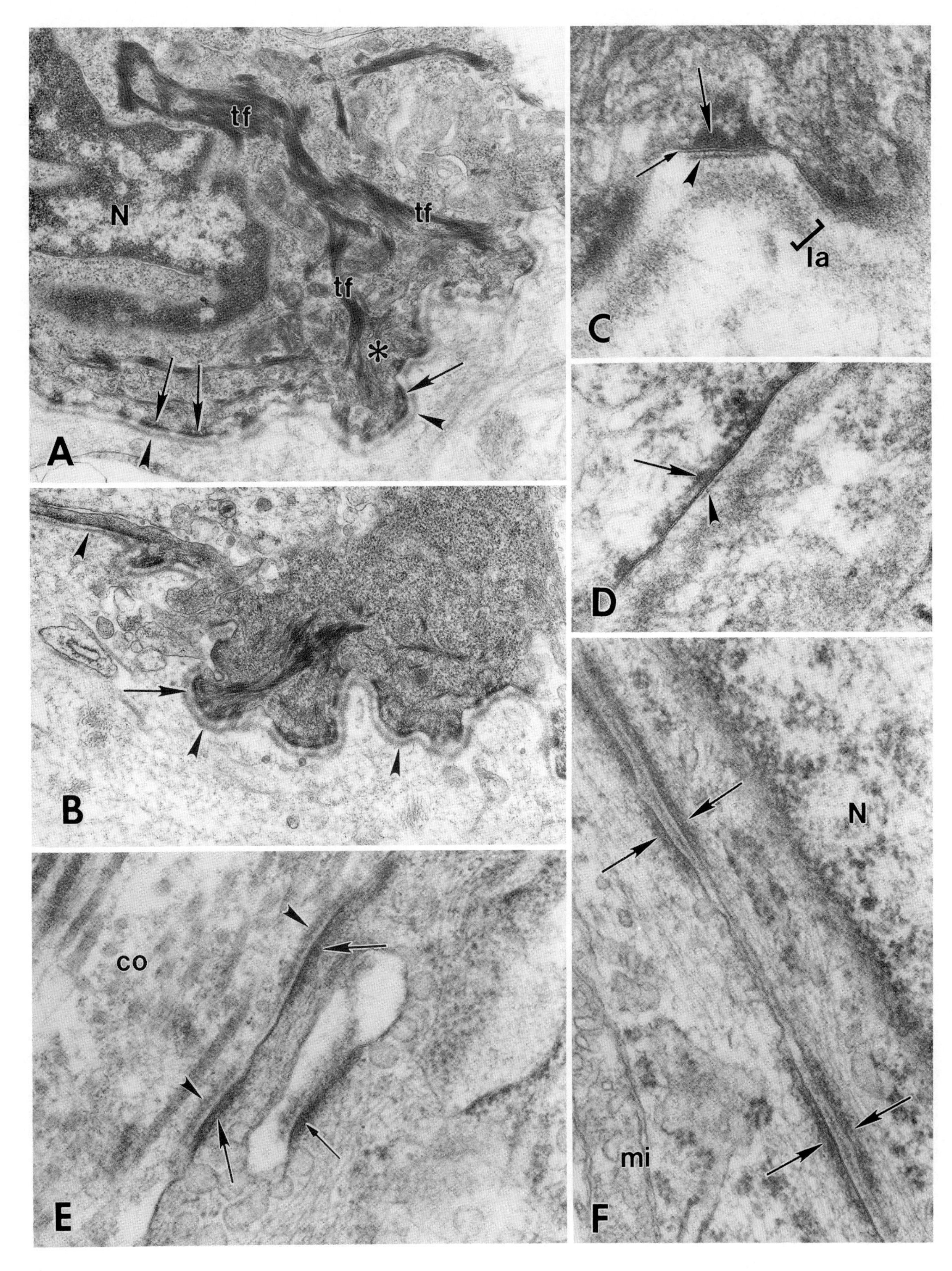
tf
tf
N
tf
*
A
B
C
la
D
co
E
N
mi
F

focal area of finely textured filamentous material on the external surface of the membrane* (Plate 66E). This extracellular material has some similarities to lamina (Chapter 27), but its relationship to lamina is completely unknown. Sometimes, several SLDs together can give the appearance of lamina.

SLDs are structurally simpler than hemidesmosomes; they are, however, sometimes confused with them. It is important to make the distinction given their different cell distributions. SLDs never have intermediate filaments contacting them, and they lack the sub-basal plate of hemidesmosomes.

* To be precise, the dense material under the membrane is itself a subplasmalemmal linear density, but in practice the term is used to include the associated lamina-like material on the cell surface.

Paired subplasmalemmal linear densities (paired SLDs) resemble two SLDs joined together such that they share the extracellular lamina-like material (Plate 66F). They are particularly common between muscle cells. Here, the material in the intercellular space probably is lamina, but in a slightly compressed form (Plate 66F). The intercellular material may vary in quantity. Sometimes it resembles a thick and fuzzy line, and the whole junction then has a superficial similarity to a desmosome. It is important, however, that such junctions are not referred to as desmosomes (as in Figure 1 in Quinonez and Simon, 1988) given the different cell specificities of paired SLDs (mesenchymal cells) and desmosomes (epithelium).

Plate 66. **A** and **B**: Cells abutting stroma. In **A**, several distinct hemidesmosomes are seen (*large arrowheads*). Although well-developed tonofibrils are present, in this section they only approach the hemidesmosomes but do not make contact. In **B**, tonofibrils make contact (*large arrow*). Observe also the well-developed continuous basal lamina (*arrowheads*) in both. **A** and **B**, × 20,500. **C**: Detail of a hemidesmosome in a small surface depression from the asterisked area in **A**. Note the amorphous submembranous material (large arrow), the sharply defined membrane (*small arrow*) and the clearly defined sub-basal plate (*arrowhead*). The basal lamina is rather fuzzy and seems to be attached to the sub-basal plate. × 110,000. **A**–**C**: Squamous epithelium of rectal polyp. **D**: Very primitive hemidesmosome represented by a small quantity of submembranous material (*arrow*) and an inconspicuous sub-basal plate (*arrowhead*). Myoepithelial cell in pleomorphic adenoma, parotid. × 78,800, **E**: Subplasmalemmal linear densities (*arrows*) at the surface of a stromal cell. One (*small arrow*) is on the surface of a small invagination. Arrowheads indicate lamina-like material outside the membrane. Spitz nevus, skin of back. × 61,100. **F**: Paired subplasmalemmal linear densities (*arrows*). Leiomyosarcoma, soft tissues of calf. × 60,000. co, collagen; la, lamina; mi, mitochondrion; N, nucleus; tf, tonofibril.

27

Lamina: Fibronexus

LAMINA

The surfaces of many cells are partially or completely covered by a material variously referred to as **basal lamina, external lamina**, or **basement membrane**. Typically, this is a 50 to 100-nm-thick layer of finely filamentous material of moderate density whose principal constituents are type IV collagen, laminin, and heparan sulfate proteoglycan. The lamina has filamentous connections with the cell membrane and with fibrils in the extracellular space; and although the lamina is often considered as belonging to a given cell, it clearly provides a connection between the cell and matrix.

The lamina is found at the interface of epithelium or epithelium-like cells or tissues and underlying stroma—for example, mesothelium, meningothelium, the granulosa cell layer, and Sertoli cells. It is also found surrounding several types of mesenchymal cell: muscle, nerve sheath, endothelium, and adipocytes. Cells lacking lamina include hematopoietic cells, fibroblasts, chondrocytes, osteoblasts, and neurons. Identifying the lamina is therefore important in diagnosing all kinds of epithelial tumors and tumors with an epithelial architecture: mesotheliomas, granulosa cell, Leydig and Sertoli cell tumors, yolk sac tumors, and ependymomas. It is also present in tumors of myogenic, schwannian, perineurial, endothelial, and adipocytic differentiation. Therefore, lamina helps to differentiate carcinoma from lymphoma and neuroblastoma; it also helps to differentiate fibroblastic proliferations, chondrosarcoma, osteosarcoma, and neuronal tumors on the one hand from malignant nerve sheath tumors, myosarcomas, angiosarcoma, and liposarcoma on the other. In the more undifferentiated examples of these tumors, however, the lamina tends to be poorly developed, fragmented, or absent.

Ultrastructure of Lamina

Typically, the **lamina** faithfully follows the contours of the cell membrane and is separated from it by a clear space of uniform width (Plates 66A, 66B, and 67A). Some examples of lamina possess **anchoring fibrils**

Plate 67. **A**: Basal lamina (*arrows*) following contour of epithelial cells abutting stroma. Normal breast. ×13,300. **B** and **C**: Lamina over smooth-muscle cells. Note attachment plaques (*arrows*), association of lamina with these plaques, and (in **C**) discontinuities (* between *arrowheads*). **B**, leiomyosarcoma, soft tissues of thigh. ×60,000; **C**, normal myometrium. ×13,000. **D**: Abnormally thick (~200 nm) and pale-staining lamina (*arrowheads*) associated with vascular smooth-muscle cell. Note irregular lamina lucida (*arrows*).Solitary focal densities are also present. Aggressive angiomyxoma, broad ligament. ×60,000. **E**: Lamina of variable thickness and with minimal lamina lucida (*arrow*). Normal perineurial cell. ×71,600. co, collagen; ecm, extracellular matrix; fd, solitary focal densities; mi, mitochondrion; N, nucleus.

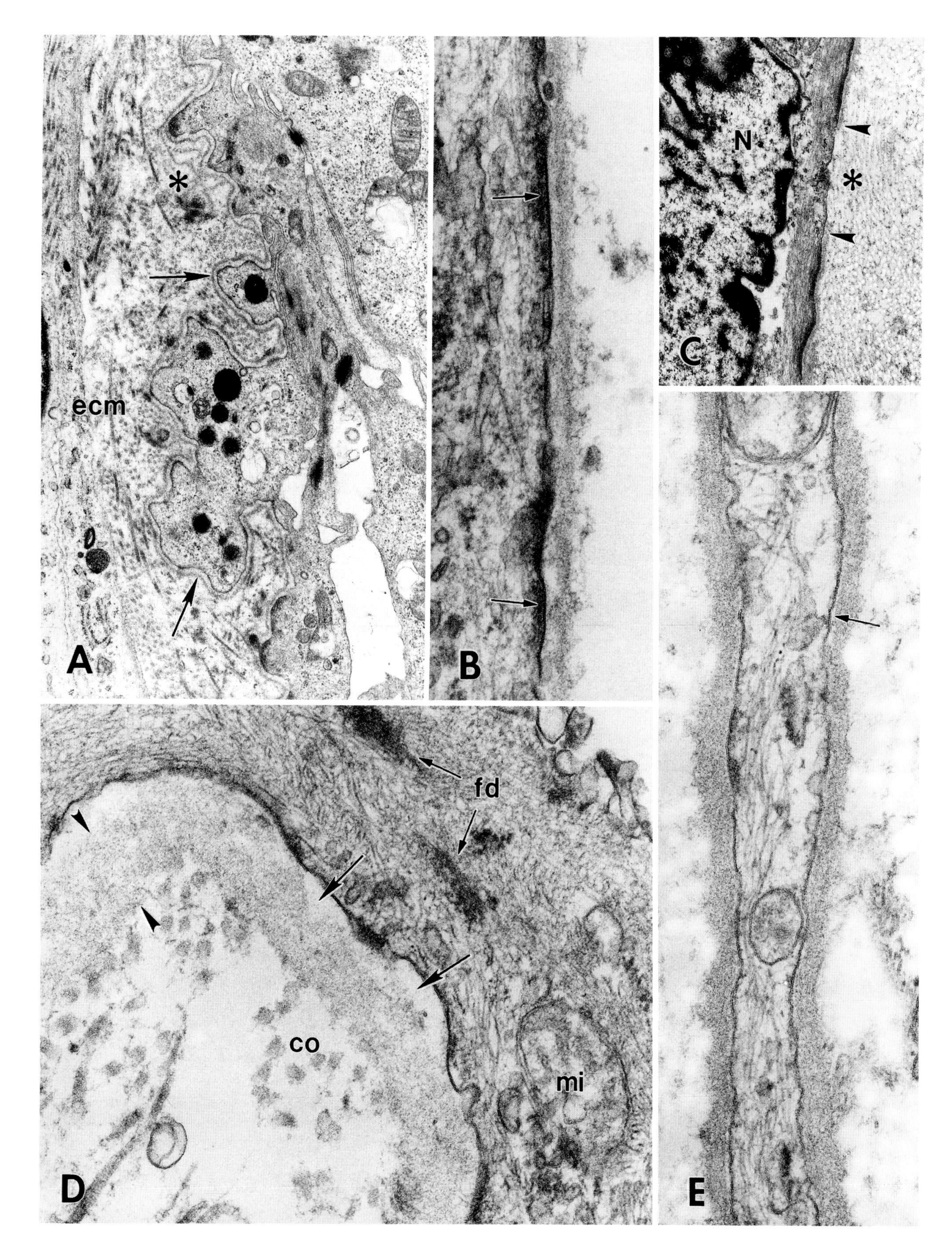
*
ecm
A
B
N
*
C
fd
co
mi
D
E

(Chapter 30) forming connections with matrix fibrils, and fine filaments link the lamina to the plasma membrane (Reale, 1984).

Terminology

The terminology in the literature associated with the lamina is often misleading or illogical; this chapter provides simplified guidelines.

Lamina which covers a basal area of a cell showing an unambiguous basal–apical polarity is referred to as **basal lamina**, and this is usually found over epithelial cells and cells in tissues with an epithelium-like architecture abutting stroma. Where a basal–apical polarity is absent—for example, in muscle, nerve sheath, and fat cells—it is clearly illogical to use **basal lamina**. Many instead use **external lamina**. All laminae are external, however, and therefore in these circumstances it is appropriate and acceptable simply to use **lamina**. The term **basement membrane** is also best avoided. This is derived from light microscopy, and it encompasses elements in addition to the lamina [e.g., anchoring fibrils (Chapter 30)]. The term is also unsuitable in electron microscopy because **membrane** has a precise significance for the lipoprotein barriers forming the plasmalemma and the boundaries of many intracellular organelles.

A refinement of terminology uses **lamina lucida** (=**lamina rara**) for the clear space next to the plasma membrane (Plate 67A), and **lamina densa** for the dense lamina itself. Usually, however, the simpler terminology of **lamina** or **basal lamina** on its own is preferred and sufficient; here, **lamina** is equivalent to **lamina densa**.

Ultrastructural Variations

In normal tissues and in well-differentiated neoplasms, lamina is usually a single-layered structure, faithfully following cell surface contours. However, in many tumors—and even in some normal cells—the lamina may appear in a number of structural variations. In muscle cells, the lamina is often associated with attachment plaques (Plate 67B) and discontinuities (Plate 67C). The lamina may be unusually thick and pale-staining (Plate 67D), while the lamina lucida may be absent or exceptionally narrow (Plate 67E). In addition, the lamina may be reticulate or multilayered (Plate 68A and 68B).

In all of these variations a degree of laminate organization persists, but some extracellular material is regarded as lamina even though it has lost this typical organization; this is **amorphous** or **granular lamina** (Plate 68C). The evidence for considering this as lamina comes from its position (sometimes obviously basal), as well as from images in which continuity with conventional lamina is observed (Stewart et al., 1981).

Sometimes, discrete amorphous masses of material are found in the extracellular matrix, and these may also be a form of lamina given their texture and occasional association with cell surfaces (Dickersin, 1987; Damjanov et al., 1989): however, they are uncharacterized in terms of protein composition. In osteogenic tumors, similar stromal bodies have also been interpreted as premineralized matrix (Papadimitriou and Drachenberg, 1994).

Lamina can survive in tissues retrieved from paraffin wax. It may, however, experience artifactual condensation and appear thinner than usual.

FIBRONEXUS

The **fibronexus (fibronexus junction)** is a cell-to-matrix specialization mediating contact between cell surfaces and extracellular matrix fibrils. In myofibroblasts, it enables the contractility generated within the cell to be transmitted to the matrix to allow overall tissue contraction, for example, in granulation tissue.

The fibronexus is found predominantly in myofibroblasts and tumors or tumor-like lesions consisting of or including myofibroblasts: myofibrosarcomas, nodular fasciitis, and malignant fibrous histiocytomas.

Ultrastructure of the Fibronexus

The fibronexus consists of two systems of filaments interacting at a localized area of plasma membrane

Plate 68. **A**: Multilayered and reticulate lamina between two tumor cells. Note abnormal melanosomes (*arrows*). Psammomatous melanotic schwannoma. ×24,000. Micrograph courtesy of J. A. Carney, M. D. reproduced from Carney (1990), with permission from Raven Press.] **B**: Straight profiles of duplicated lamina in a Call–Exner body (*arrows*). Granulosa cell tumor, metastatic to liver. ×13,000. **C**: Granular lamina (*), some of which is closely associated with cell membrane (*arrows*). Chondroid syringoma, palm. ×40,400. if, intermediate filaments; mi, mitochondria; N, nucleus.

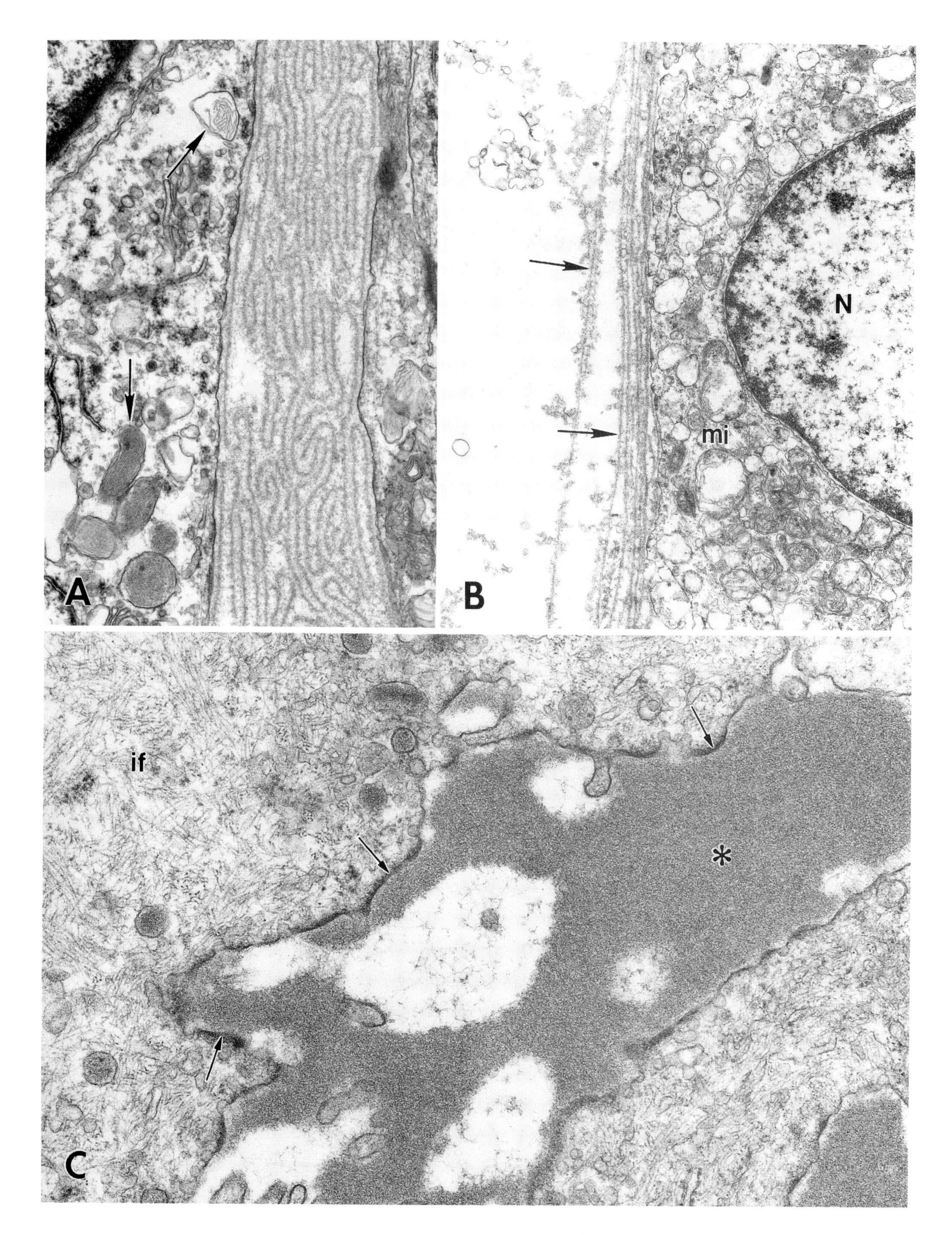
A
B
N
mi
if
*
C

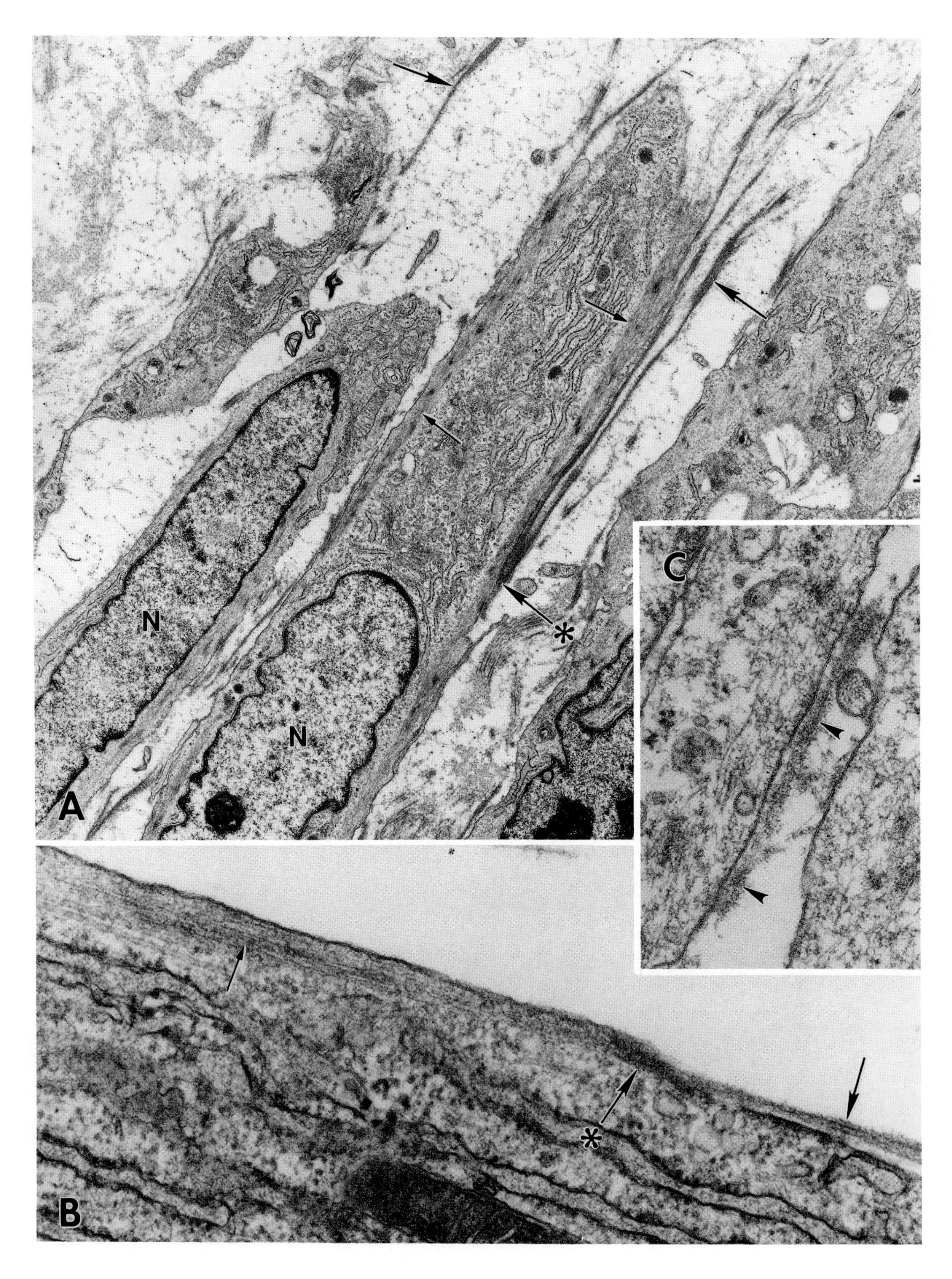
N
N
A
C
B

marked by a subplasmalemmal plaque (Plate 69). Peripheral intracellular smooth-muscle myofilaments converge on this plaque in a bundle. In the extracellular space at this site there are fine fibronectin filaments which bundle together to form the **fibronectin fibril**; this diverges away from the plaque and into the stroma (Plate 69A). The filamentous substructure of the fibronectin fibril is only observed at high magnification (Eyden, 1993). The point at which intracellular myofilaments and extracellular fibronectin filaments converge on the plaque and plasma membrane is the **fibronexus** or **fibronexus junction**. Sometimes the fibronexus is sited at an obvious "shelf" of the cell surface (Plate 69B). When the fibronectin fibril is highly developed and conspicuous, it serves as a marker for the fibronexus as a whole.

The literature contains several older terms for the fibronectin fibril which are best avoided: *microtendon, anchoring filament*, and *microtendinous prolongation of basement membrane.*

The fibronectin fibril differs from lamina in diverging from the cell surface, while lamina follows surface membrane contours. The fibronectin fibril is also more rigid-looking and has a distinct longitudinal filamentous substructure which is absent from lamina. It remains the case, however, that distinguishing poorly developed lamina from a poorly developed fibronectin fibril can be difficult (Plate 69C); unambiguous light microscope immunostaining or immunoelectron microscopy may be required to make the distinction. Strong fibronectin staining will suggest a fibronectin fibril; weak fibronectin staining in combination with convincing laminin and type IV collagen staining will suggest lamina.

Collagen and oxytalan fibrils (Chapter 30) can be found juxtaposed to the plasma membrane and should not be confused with the fibronectin fibril.

Plate 69. **A**: Desmoplastic myofibroblasts showing intracellular myofilaments (*small arrows*) and extracellular fibronectin fibrils (*large arrows*). The fibronexus is their point of convergence (*asterisked arrow*). Primary invasive squamous cell carcinoma, neck. ×8100. **B**: Normal lung fibroblast in culture showing rather poorly developed myofilaments (*small arrow*), very straight filaments forming a fibronectin fibril (*large arrow*), and the shelf-like point of convergence (*asterisked arrow*). ×60,800. **C**: Vaguely defined material on the cell surface (*arrowheads*). This is difficult to identify as lamina or as a fibronectin fibril. Spindle cell sarcoma NOS, subcutis forearm. ×61,000. N, nucleus.

28

Processes: Lumina

PROCESSES

Processes are projections of the cell surface maximizing interaction with the extracellular environment. The best known function associated with cell processes is absorption in intestinal and other luminal microvilli. Tumor cells show a diversity of cell processes, some of which have been given distinctive names, partly based of their appearances in thin sections (e.g., *lobopodia, filopodia, vermipodia*). These names do not add significantly to understanding the nature or function of these processes, and in some instances the name is a poor reflection of the real three-dimensional structure; some of the filopodia of macrophages, for example, are probably flaps or ruffles. The term **microvillus** is classically applied to gastrointestinal tract cell processes and can also be applied to all glandular or apical–epithelial processes (Plate 70A and 70B). In other situations, one should simply use the term **cell process**, with or without a prefix denoting the cell type—for example, *mast cell process, Schwann cell process.*

Cell processes have characteristic appearances which contribute to the identification of several types of cell found as reactive elements in tumors; they are also diagnostically important in a variety of glandular epithelial tumors, as well as in tumors showing glandular epithelial architecture. The following are the main areas of importance: adenomas (Plate 19), adenocarcinomas (Plate 61A), adenoid cystic carcinomas, mesotheliomas (Plate 70C), biphasic synovial sarcomas (Plate 65A), Wilms' tumor, endodermal sinus tumors (Plate 21E); macrophages (Plate 16A), multinucleated giant cells (Borg-Grech et al., 1987), mast cells (Plate 14B), leukemias and especially hairy cell leukemias (Plate 72A); the anemone cell tumors (lymphomas, carcinomas, neuroendocrine tumors, ependymoma) (Plate 71C). Mesenchymal or partially mesenchymal cells such as breast stroma fibroblasts (Plate 72B), myoepithelial cells, dermatofibrosarcoma protuberans, and hemangiopericytomas often have long, slender, polar processes; Schwann cell tumors (Plate 72D) and myoepithelial cells (Watson et al., 1988) have processes coated in lamina; chondroblasts often have scalloped cell surfaces with fine processes (Erlandson and Huvos, 1974); coarse cell processes full of organelles (especially intermediate filaments and microtubules) are

Plate 70. **A** and **B**: Normal duodenum showing microvilli (obliquely sectioned in **B**) with actin-filament rootlets (*small arrows*). **B** shows rootlets interacting with the terminal web (*large arrows*). Both ×20,108. **C**: Long, slender, and smooth microvilli (*arrows*). Peritoneal epithelial mesothelioma, ×39,600. **D**: Short stubby microvilli (*arrows*) and glycocalyceal bodies (*) in an intracytoplasmic lumen. Urinary bladder adenocarcinoma. ×38,900. **E**: Granular glycocalyx. Normal nasal mucosa. ×48,800. **F**: Glycocalyx consisting of filaments (*arrowhead*) and granules (*arrow*). Normal breast. ×61,100. lu, lumen; mv, microvilli.

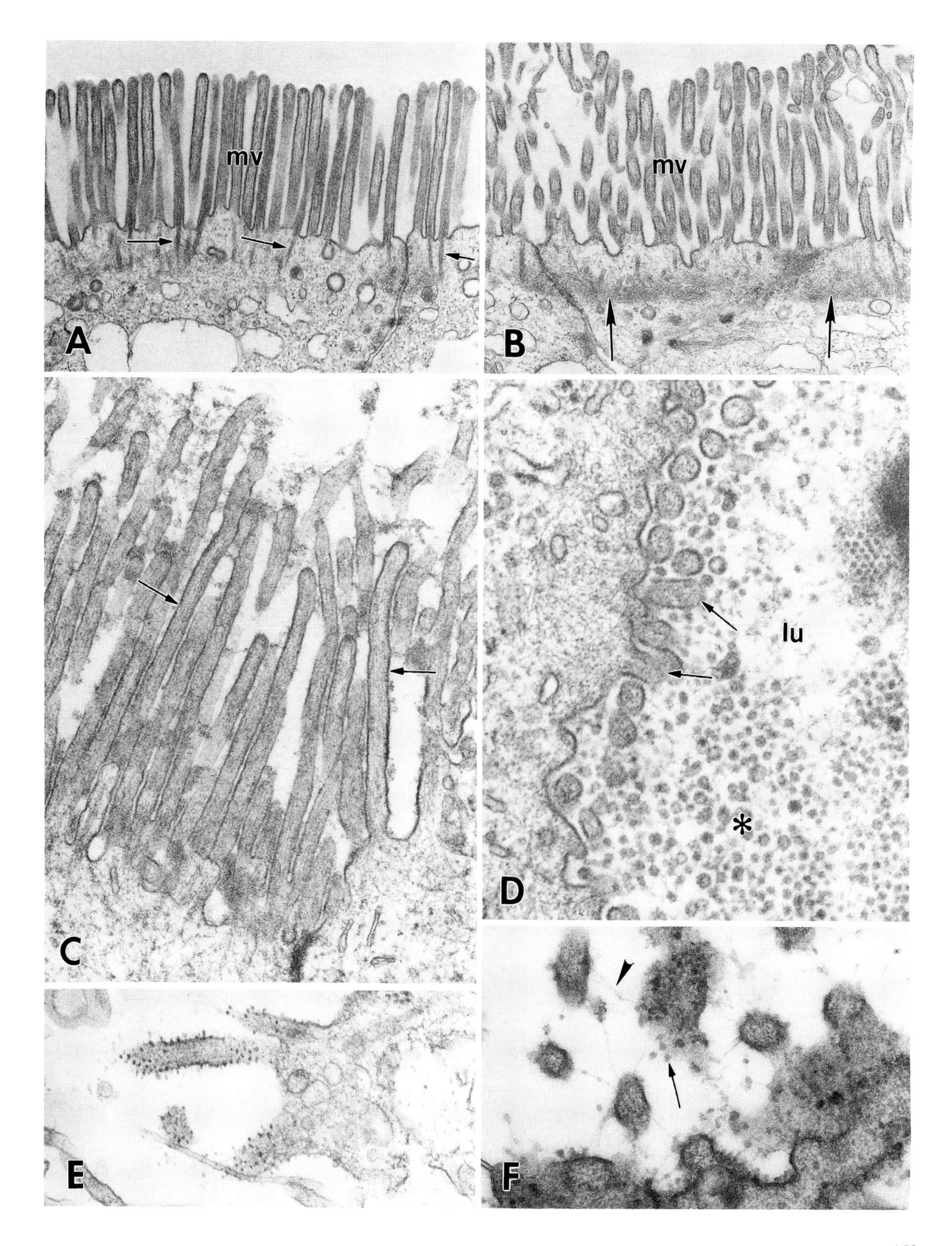
mv
A
mv
B
C
lu
*
D
E
F

found in central nervous system and neuronal tumors such as gastrointestinal autonomic nerve tumors, while longer, more slender processes containing microtubules are typical of neuronal tumors such as neuroblastoma (Plate 62); processes compressed between cells appear as interdigitations and are found in many cells and tumors: (i) Langerhans cell granulomatosis (Plate 33A); (ii) leukemic hairy cells compressed with erythrocytes in the spleen; (iii) macrophages compressed between each other and other cells; (iv) lateral cell surfaces of glandular or squamous epithelium (Plate 71A); (v) meningioma, epithelioid sarcoma, interdigitating reticulum cell sarcoma; (vi) between endothelial cells in vessels (Plate 32A) (the luminal surface of endothelium also has fine processes which look finger-like but may be ruffles or flaps in three dimensions); basal processes forming the *basal labyrinth* occur in renal epithelial cells and neoplasms (Storkel et al., 1989).

Ultrastructure of Microvilli and Cell Processes

The microvilli covering the free surfaces of cells of internal epithelium and their neoplastic counterparts are slender and finger-like, are collectively orientated at right angles to the cell surface, sometimes possess a granular or filamentous covering, and contain straight **actin filaments** (Plate 70A). These filaments provide cytoskeletal support and when numerous form a bundle called the **actin-filament core**. This sometimes extends as the **actin-filament rootlet** (Plate 70A) into the peripheral cytoplasm; here, it joins the **terminal web** of non-muscle-type actin filaments (Plate 70B). The older term *microfilament core* should be avoided. Actin-filament cores are classically associated with intestinal microvilli, but a few primary adenocarcinomas in non-gastrointestinal tract sites (e.g., lung) also possess these cores and rootlets.

Some microvilli have a covering of small granules and/or filaments called the **glycocalyx** (Plate 70E and 70F). Larger granules or vesicles are referred to as **glycocalyceal bodies** (Plate 70D), and these too are mostly though not exclusively associated with intestinal adenocarcinomas.

Microvilli vary in length and diameter, and they vary in concentration on the cell surface. While normal intestinal microvilli are fairly long and narrow, paradoxically microvilli of adenocarcinomas and cystadenomas, for example, are typically quite short and blunt (Plate 70D): mesothelial microvilli, by contrast, show a range of length but can be very much longer than typical adenocarcinomatous microvilli (Plate 70C). Differences in the length-to-diameter ratio have been used to differentiate some mesotheliomas from adenocarcinomas; a value over 15 is highly suggestive of mesothelioma (Henderson et al., 1992).

LUMINA

Microvilli are found in large numbers in association with intercellular and intracellular lumina; lumina, junctional complexes, and microvilli in fact help to define one another. These three features collectively represent the most important ultrastructural marker of glandular epithelial differentiation in normal and neoplastic tissues.

Views of intercellular lumina are shown in Plates 19, 21E, 40B, 52C, and 65A; vascular lumina are illustrated in Plates 27A and 32A.

Intercellular lumina are defined by the presence of several apical cell surfaces contributing to the luminal surface; **intracellular lumina** are found entirely within a single cell (Plate 71A).

Like intercellular lumina, intracellular lumina are indicative of glandular epithelial cell differentiation (Plate 71A); they are well known in lobular and other breast carcinomas, for example, but structures resembling intracytoplasmic lumina can be found in a wide range of tumor cells not considered as primarily showing glandular epithelial differentiation: paragangliomas, medullary thyroid carcinomas, chordomas, malignant melanomas, squamous cell carcinomas, and mesotheliomas.

The following features associated with lumina are worth noting.

1. Some intercellular lumina are too small to be easily recognized by light microscopy (Plate 71B); these are

Plate 71. **A**: Intracytoplasmic lumen. Note also interdigitations (*arrows*). Normal breast. × 10,100. **B**: An almost closed intercellular lumen. Only some of the apical surface is readily observed (*arrow*). Microvilli (*arrowheads*) are slightly irregular. The curved arrow indicates either an invagination of the main lumen or a separate intracytoplasmic lumen. The main intercellular lumen is demarcated by junctional complexes consisting of intermediate junctions and probably tight junctions, although the latter at this magnification are not quite resolved. One junctional complex has a desmosome and is therefore complete. Carcinosarcoma, oral cavity. × 38,900. **C**: A large cytoplasmic process (*) producing and surrounded by large numbers of slender and elongate cell processes (*arrows*) completely filling out the intercellular area, typical of the anemone cell appearance. Hodgkin's disease, axillary node. × 11,400. d, desmosome; j, intermediate junction; il, intracytoplasmic lumen.

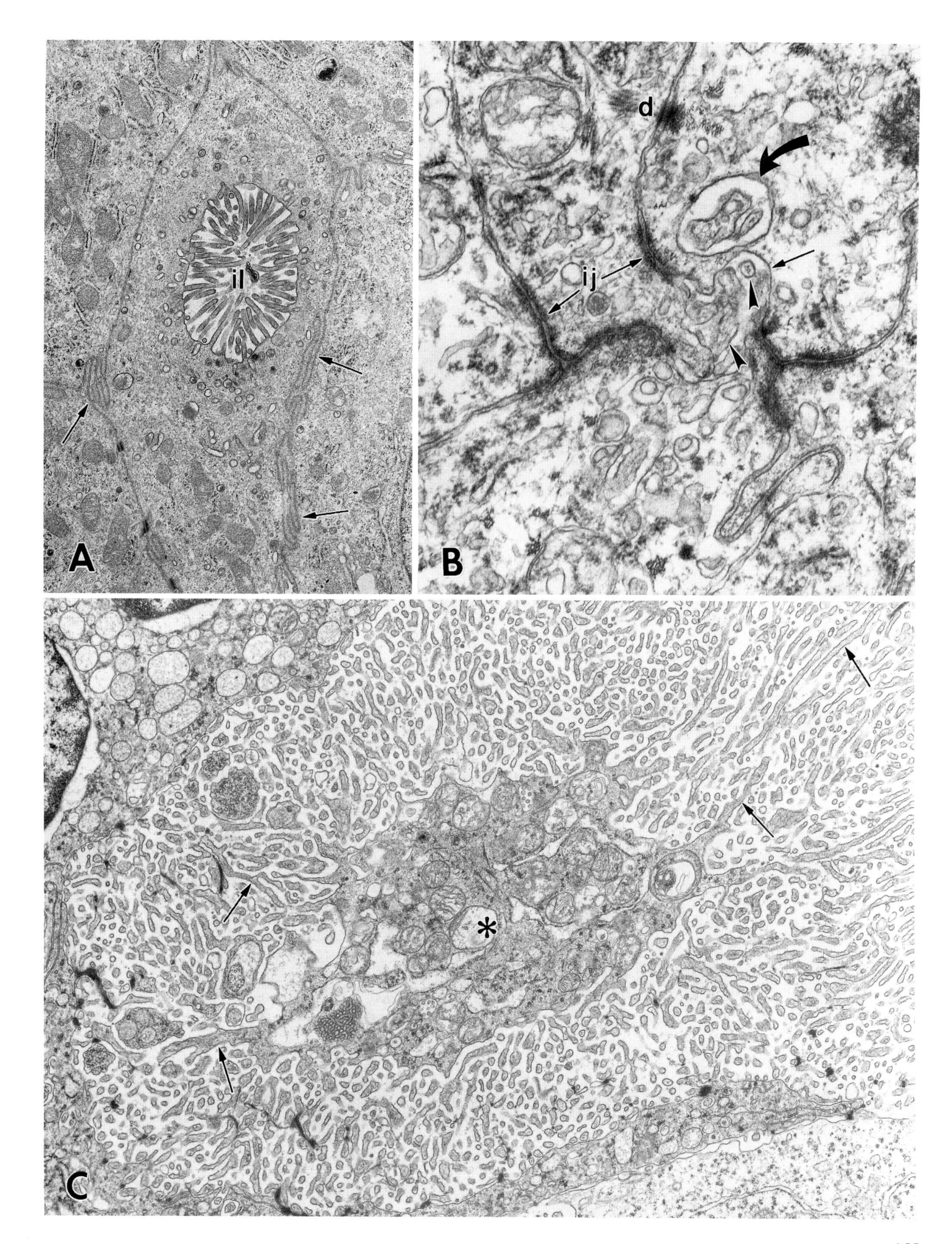
il
A
d
ij
B
*
C

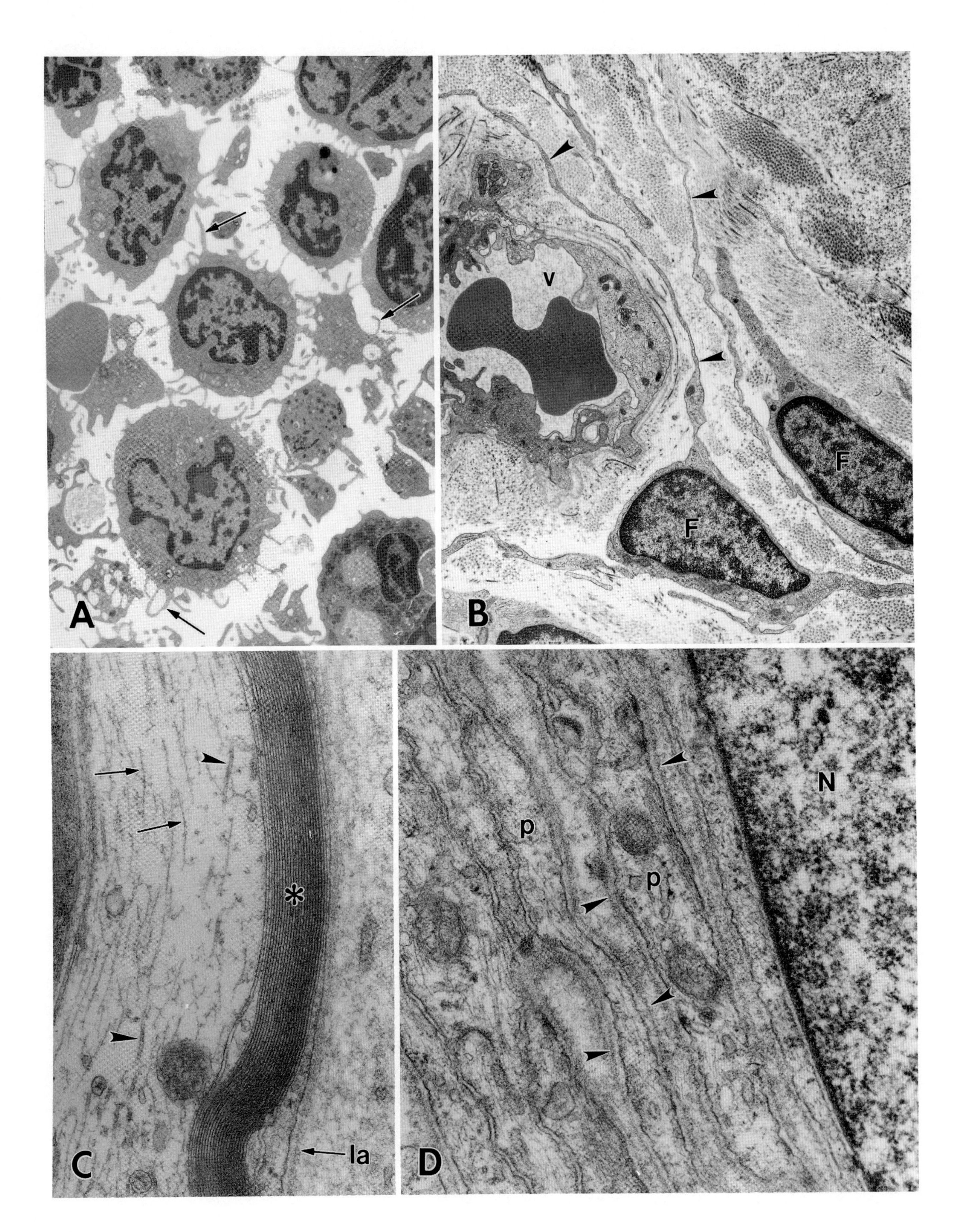
A
B
V
F
F
C
*
la
D
p
p
N

microlumina and their identification is important, for example, in revising a diagnosis of carcinoma for a tumor consisting of sheets of poorly differentiated epithelial cells to adenocarcinoma.

2. The luminal space can be occupied by a variety of structures quite apart from microvilli: cilia, more or less homogeneously dense secretion (Plate 40B), exocrine granule cores, crystals, cells and cell debris, psammoma bodies and other forms of calcification.
3. Ideally, lumina should be defined in terms of not only an intercellular space but one demarcated by junctions, as in Plate 71B; otherwise, one may be looking at nothing more than a localized intercellular space of no diagnostic value.

Plate 72. **A**: Processes (*arrows*) in hairy cell leukemia. Buffy layer from peripheral blood. ×5200. **B**: Slender elongate processes (*arrowheads*) in intralobular stromal fibroblasts. Normal breast. ×9000. **C**: Schwann cell myelin sheath (*) associated with a neuronal process containing neurofilaments (*arrows*) and microtubules (*arrowheads*). Note Schwann cell lamina. Ganglioneuroma, metastatic to cervical node. ×39,600. **D**: Schwann cell processes coated in lamina (*arrowheads*). The processes are considered to represent attempted myelin sheath formation. Benign schwannoma, soft tissues of chest wall. ×38,900. F, fibroblast; la, lamina; N, nucleus; p, process; v, vessel.

29

Crystals

Geometrical intracellular inclusions and matrix structures with an ordered substructure are referred to as **crystals** or **crystalloids**. Some are membrane-bound Golgi secretions (Plate 73B); others are free in the cytoplasm (Plate 73D), nucleus, or extracellular space. Most are proteinaceous.

The most commonly encountered crystals and crystal-like structures are: immunoglobulin within rough endoplasmic reticulum in lymphoid (Plate 73A) and especially plasma cell proliferations; rhomboidal renin crystals in juxtaglomerular cell tumors (Plate 73B); rhomboidal crystals in alveolar soft-part sarcomas (Plate 73C); Reinke crystals in Leydig cells and Leydig cell tumors (Plate 73D and 73E); cystine inclusions in macrophages in renal allografts (Spear et al., 1989); intraluminal crystalloids in struma ovarii and adenosquamous carcinomas (Ro et al., 1991; Van Hoeven et al., 1993); Charcot–Leyden crystals, characterized by hexagonal profiles and Charcot–Leyden crystal protein, in diverse clinical settings (Carson et al., 1992). Crystals also occur in angiomyolipomas (Mukai et al., 1992), human hepatocytes in association with alcoholic liver disease (Schaff et al., 1990), Sertoli cells (Nistal et al., 1982), and a subependymal giant cell tumor (Jay et al., 1993); *serpigenous* inclusions occur in rough endoplasmic reticulum (Chapter 3).

Plate 73. **A**: Immunoglobulin in rough endoplasmic reticulum (*arrows*), showing crystalline structure (*arrow*, **inset**). Diffuse large B-cell lymphoma, diverse nodes. [Reproduced from Eyden (1994), with permission from Gustav Fischer Verlag.] ×61,000 (**inset** ×70,000) **B**: Membrane-bound rhomboidal crystals (*arrows*). Primary juxtaglomerular cell tumor. [Micrograph courtesy of Dr. R. Kodet (Prague). Reproduced from Kodet (1994), with permission from Lippincott–Raven Publishers.] **C**: Rhomboidal crystals with a bidirectional periodicity. Alveolar soft-part sarcoma, intramuscular mass, thigh. ×73,000. [Micrograph courtesy of Birgitta Carlen (Lund, Sweden).] **D**: Reinke crystals (*). Leydig cell tumor, ovary. ×6500. [Micrograph courtesy of Professor N. Schnoy (Berlin). Reproduced from Schnoy (1982), with permission from Springer-Verlag.] **E**: Reinke crystal showing linear substructure (*arrows*). Leydig cell tumor, testis. ×28,000. N, nucleus; Nu, nucleolus.

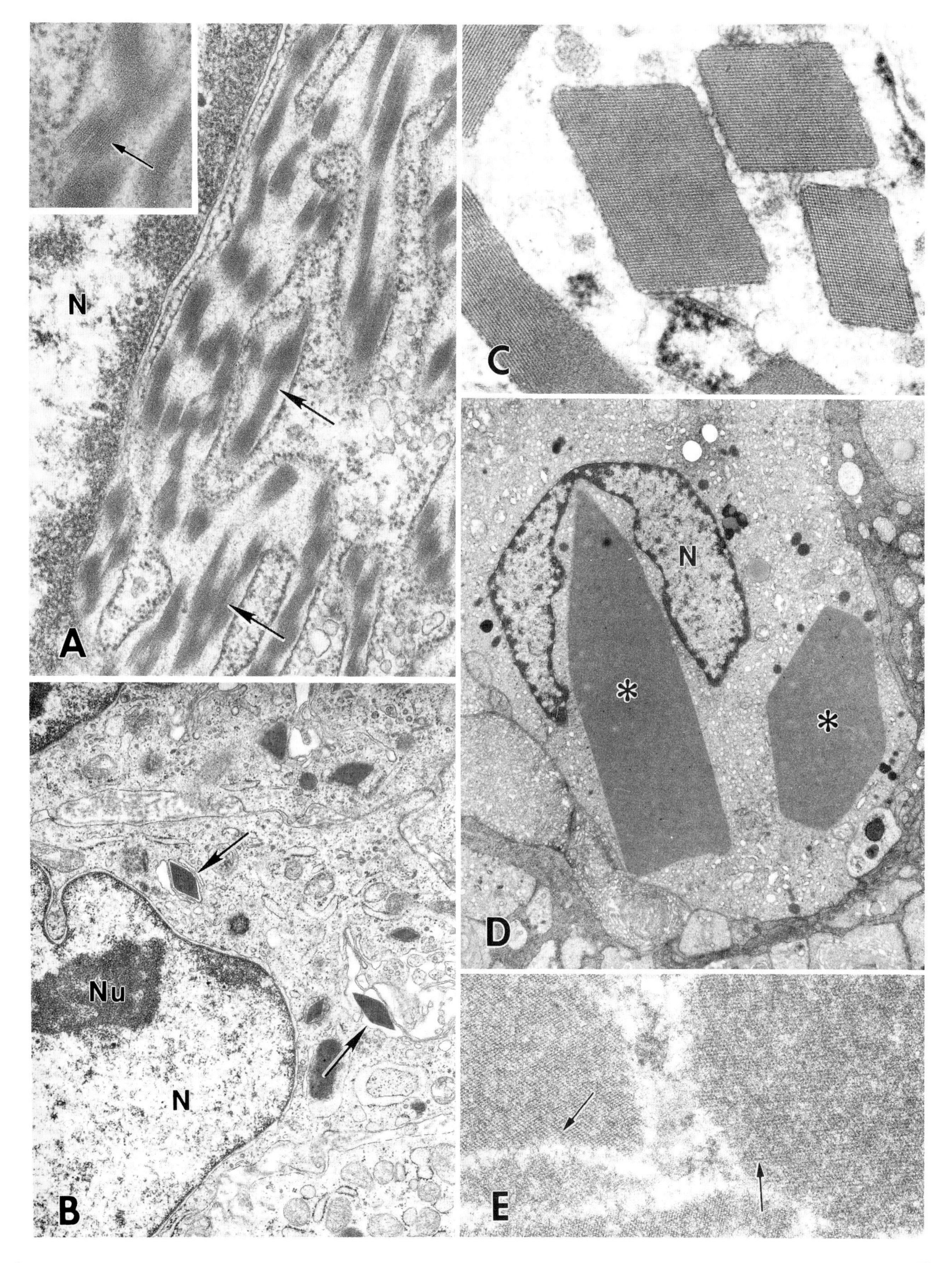
N
A
C
N
D
Nu
N
B
E

30

The Extracellular Matrix

The space outside the plasmalemma is referred to as the **extracellular space**. When this is clearly between cells and essentially structureless (as between the lateral membranes of glandular or squamous epithelium), it can also be referred to as the *intercellular* space. The collective mass of fibrils and filaments in the extracellular space is referred to as **extracellular matrix (ECM)**, and often this is considered synonymous with the term **stroma**.

The main ultrastructurally identifiable matrix components are: fibrillar collagen, elastin, amyloid, fibrin, and proteoglycans. Microorganisms and foci of calcification may also be encountered in the ECM. All of these components can survive paraffin wax embedding.

FIBRILLAR COLLAGEN

Fibrillar collagen is probably the most abundant protein of the ECM, and at least some is found almost ubiquitously in tumors. When assembled in groups, collagen fibrils constitute a **collagen fiber** (Plate 74A), and these are of a size easily seen by light microscopy. Collagen fibrils typically have a rounded cross-sectional profile (Plate 74B) and a distinctive cross-striated periodicity of ~50 nm (Plate 74C). Fibrils with these features are composed of collagens type I and III.

Collagen fibrils vary in structure and diameter. In normal tissues and some tumors, they have a fairly uniform diameter (Plate 74B), typically ~50–100 nm. Thinner fibrils, down to ~20 nm, are still recognizable as collagen if they retain the typical 50 nm cross-striations. Fibrils between 20 and 100 nm are referred to as **native collagen fibrils**, and this is considered to be the normal size range.

Collagen fibrils can vary in diameter from this normal range up to about 1 μm, and these are considered as abnormally thick fibrils. Many of these abnormally thick fibrils have an irregular outline (Plate 74D) and have sometimes been referred to as **collagen flowers**. Very-large-diameter fibrils (approaching 1 μm)

Plate 74. **A**: A large bundle of collagen fibrils forming a fiber (*). Note the nearby lymphocyte for size (L) (4 μm across). Intralobular stroma of normal breast. ×12,100. **B**: Transversely sectioned collagen fibrils of regular profile and fairly uniform size (*arrows*). The filamentous material (*arrowheads*) around the fibrils is either proteoglycan, glycosaminoglycan, or other collagen molecules. Fibrothecoma, ovary. ×39,600. **C**: Detail of a single collagen fibril showing cross-striations and adherent filamentous material (*arrowheads*). Stroma of cervical polyp. ×92,300. **D**: Collagen fibrils of regular (*small arrow*) and irregular (*large arrows*) cross-sectional profile. Same tumor as in **B**. ×39,600. **E**: Pale-staining collagen fibrils (*arrows*) embedded in dense amorphous matrix. Hodgkin's disease, axillary node. ×13,300.

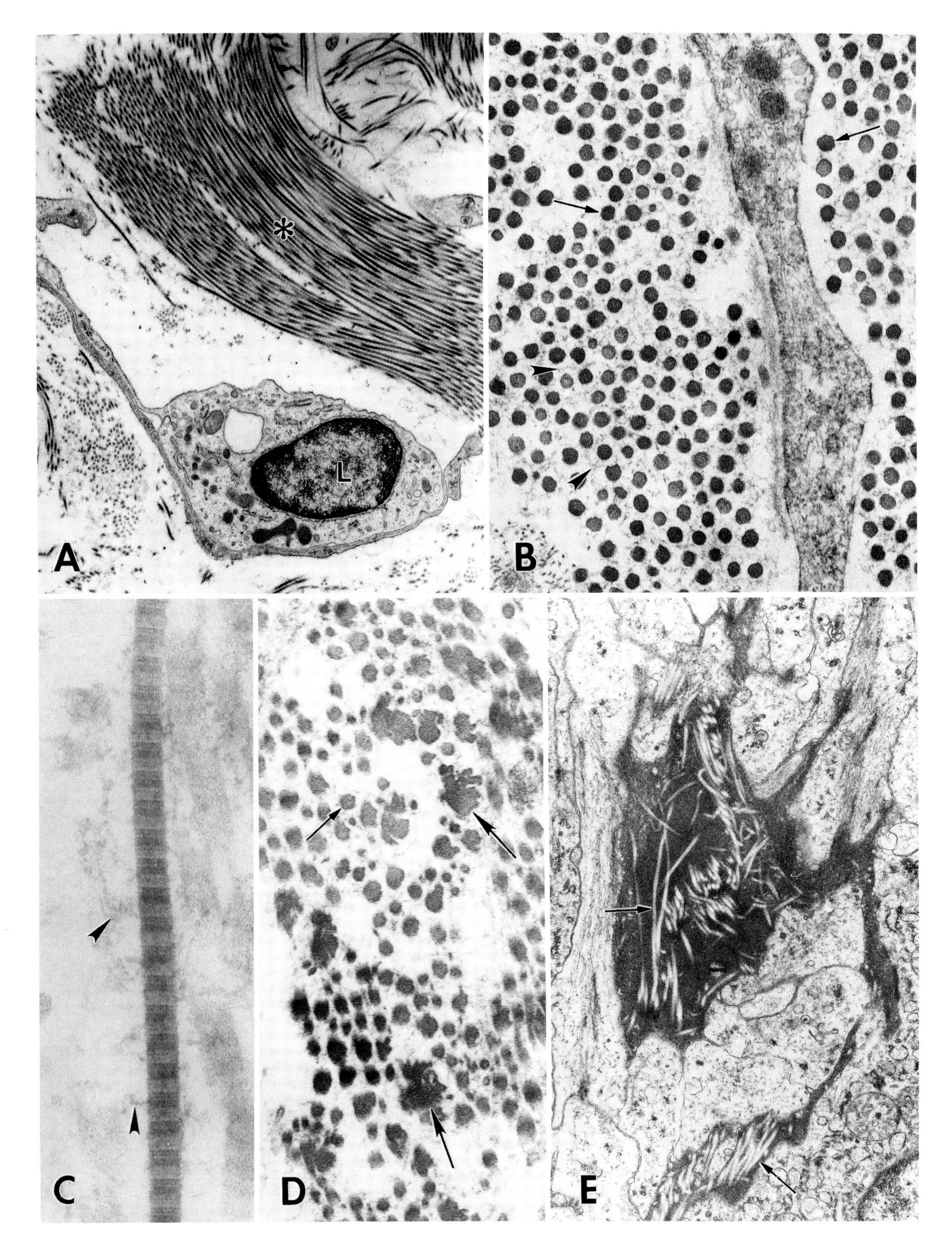

*
L
A
B
C
D
E

have been described as **amianthoid** following use of the term by Hough et al. (1973) for bundles of abnormally large collagen fibrils equivalent to the silvery or asbestos-like threads seen on the cut surfaces of costal cartilage. The term *amianthoid fibril* or *fiber* has also been used for a number of light microscopically observed collagen fibril assemblies in tumors, some of stellate configuration; often, however, the fibril diameter in these studies is in the normal range (discussed in Eyden et al., 1996). Therefore, the term *amianthoid* should only be used for those collagen aggregates where electron microscopy reveals fibrils approaching 1 μm in diameter.

True amianthoid fibers (i.e., consisting of collagen fibrils of about 1 μm in diameter) have been seen in chondrosarcoma, malignant schwannoma, and synovial sarcoma. Abnormally thick collagen fibrils (100 to more than 500 nm) are of more common occurrence (Plate 74D) and appear to predominate in mesenchymal tumors.

Collagen fibrils can show a wide range of staining density. Some collagen fibrils are pale-staining, and their outline is accentuated by the dense matrix surrounding them (Plates 32C, 74E, and 77C). See Plates 30D and 32D for densely staining collagen fibrils.

Structural Variants of Collagen

The following further structural variants of collagen exist: fibrous long-spacing collagen, anchoring fibrils, and skeinoid fibers. Of these, only the anchoring fibril occurs as a normal structure.

Fibrous Long-Spacing Collagen

Fibrous long-spacing collagen (FLS collagen) has a lateral cross-banding which is coarser and has a periodicity greater than the 50 nm typical of normal collagen fibrils. Two broad types occur. **Compact FLS collagen** consists of well-circumscribed fibrils, sometimes fusiform, found either free in the matrix or associated with cell surfaces. They possess dense lateral bands separated by 100–150 nm and joined by longitudinal fine filaments (Plate 75A and 75B). They have been traditionally referred to as *Luse bodies.* **Dispersed FLS collagen** also has lateral bands, but with a somewhat reduced period of 80–100 nm. Often also, the fine filaments between the bands are less conspicuous and the overall fibril is less sharply circumscribed compared with compact FLS collagen (Plate 75C).

Compact FLS collagen is found predominantly in benign and malignant schwannomas and in traumatic and acoustic neuromas. Dispersed FLS collagen has a wider distribution: lymphomas, Hodgkin's disease (Plate 75C), malignant melanomas, atypical fibroxanthomas, and a variety of neuroendocrine tumors and carcinomas.

Anchoring Fibrils

Anchoring fibrils are short, often curving fibrils attached to the stromally orientated surface of basal lamina (Plate 76A). They are up to ~300 nm long, are cross-striated, and are known to consist of type VII collagen. They appear to function in the anchoring of cells to matrix. The term *anchoring filament* should not be used for these fibrils; this has a prior usage for the hollow oxytalan filaments ("microfibrils") associated with the surface of lymphatic endothelium (Leak and Burke, 1968).

Anchoring fibrils are found associated with the lamina in basal epidermal cells, mammary epithelial cells, oral mucosal epithelium, and occasional carcinomas at these sites; they are also found associated with various urogenital tract epithelia (for example, prostate), but not with the lamina of mesenchymal cells or endothelium (Eyden and Ferguson, 1990).

Skeinoid Fibers

Skeinoid fibers are circumscribed masses of collagenous fibrils assembled in such a way as to resemble balls of wool—hence their name. They are mostly rounded or oval in outline (Plate 76B) but can be more elongate or irregular when compressed into restricted spaces between cells.

Skeinoid fibers have been found in tumors showing nerve sheath or neuronal differentiation: small intestinal stromal

Plate 75. **A, B**: Compact FLS collagen. In **A**, fibrils are mostly free in the matrix (*arrows*). **B** shows detail of the asterisked fibril in **A**. Spindle cell sarcoma NOS, breast. **A**, ×5300; **B**, ×39,600. (**B** reproduced from Dardick I. *Handbook of Diagnostic Electron Microscopy for Pathologists-in-Training*, New York, Igaku-Shoin, 1996, with permission.) **C**: Dispersed FLS collagen forming broad expanses (*arrows*) between several cells or cell processes. Hodgkin's disease, axillary node. ×8500.

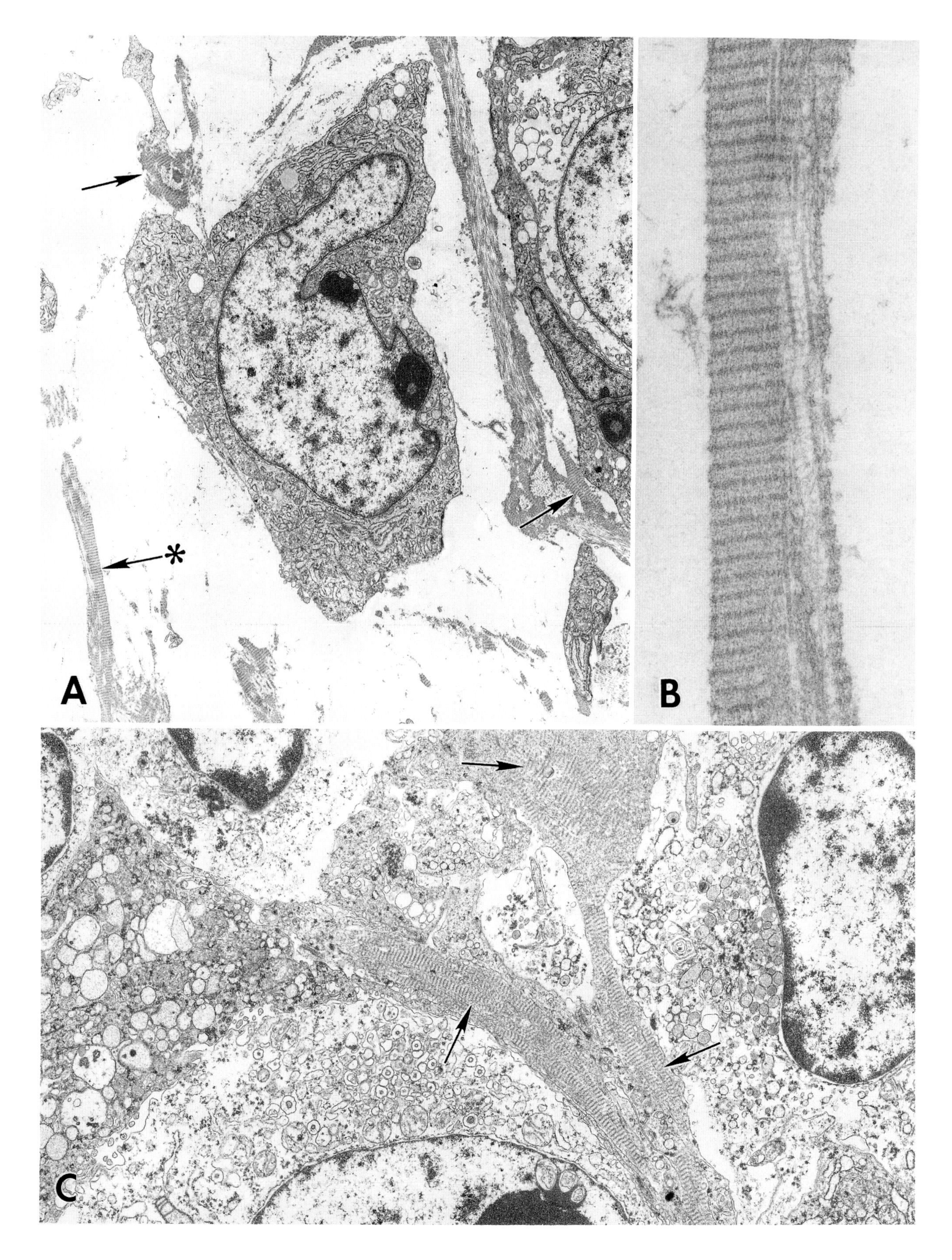
A
*
B
C

tumors associated with neurofibromatosis, gastrointestinal autonomic nerve tumors, schwannomas, and neurofibromas.

ELASTIC FIBERS

Mature **elastic fibers** are found in pliable connective tissues. They consist of (a) an amorphous component—the protein, **elastin**—containing discrete linear densities and (b) a covering of hollow filaments. The amorphous elastin component is very finely textured and usually has only a moderate density, often appearing gray in electron micrographs (Plate 77A). On the surface of the elastic fiber are tubular filaments, ~10–12 nm in diameter. These have traditionally been referred to as *glycoprotein microfibrils*. The term *microfibril*, however, is obsolete, and it is preferable to call these structures **elastin-associated filaments**.

Abnormal elastin fibers have been described in elastofibroma (Fukuda et al., 1987; Kumaratilake et al., 1991).

Oxytalan Fibrils

Filaments identical morphologically to elastin-associated filaments can also be found in aggregates on their own—that is, in the absence of the amorphous elastin component. These aggregates are **oxytalan fibrils**, and the individual filaments are **oxytalan filaments**. Because the oxytalan filaments appear to be identical to those found on the surface of mature elastic fibers, oxytalan fibrils are sometimes also referred to as **pre-elastin**.

Oxytalan fibrils are very widely distributed both in normal tissues and in tumors. In normal tissues they are typically found as small aggregates of filaments running in the same direction, often embedded in collagen. In tumors they vary considerably in size.

AMYLOID

Amyloid forms dispersed, spheroidal, stellate, or more irregular masses of material in the ECM which at low magnification appear amorphous and of moderate density (Plate 78A). Under high power, the masses consist of fibrils in disordered orientations (Plate 78B). At very high magnifications (~ × 100,000), the fibrils are seen to be hollow (Plate 78B). Amyloid fibrils are believed to consist of filaments (Tan and Pepys, 1994), although these are not identifiable in conventional electron microscopy preparations. Diameters quoted vary from 7 to 13 nm, with an occasional unusually high value (~30 nm). Amyloid fibrils should also not be confused with elastin-associated filaments; these are of similar diameter and are also hollow, but they occur in much more orientated aggregates (Plate 77C).

Amyloid occurs very widely in non-neoplastic conditions, but tumors containing significant amount of amyloid include: Hodgkin's disease, non-Hodgkin lymphomas, myelomas, hairy cell leukemias, renal carcinomas, carcinomas of gut, lung and urogenital tract, basal cell carcinomas, medullary thyroid carcinomas, pancreatic islet cell tumors, and pituitary adenomas.

FIBRIN

Fibrin is a common ECM component formed following hemorrhage. Under low magnification, fibrin appears as dense and straight or slightly curving fibrils (Plate 79A). At high power, fibrin may reveal both a longitudinal filamentous substructure and a transverse periodicity of ~13–20 nm (Plate 79B).

PROTEOGLYCANS

Proteoglycans are protein–polysaccharide molecules with an immensely branched molecular structure and water-retaining properties; they therefore confer a considerable degree of hydration to the ECM.

They are widely distributed in dermis and other connective tissues, but they are especially prominent in cartilage and the extracellular matrix of chondrosarcoma. They are

Plate 76. **A**: Anchoring fibrils (*large arrows*) in contact with basal lamina (*small arrow*) associated with normal breast epithelium. × 35,200. Micrograph courtesy of Dr. Carolyn Jones (Manchester, England). **B**: Several skeinoid fibers (*) in close association with tumor cells. × 6700. **C**: Detail of skeinoid fiber showing individual collagenous fibrils with a periodicity of ~40–50 nm. The fibrils are fairly closely packed, some appearing in long section (*arrowheads*), others in cross section (*). × 60,800. **B** and **C**: Gastrointestinal autonomic nerve tumors, primary in the small intestine and metastatic in the omentum, respectively.

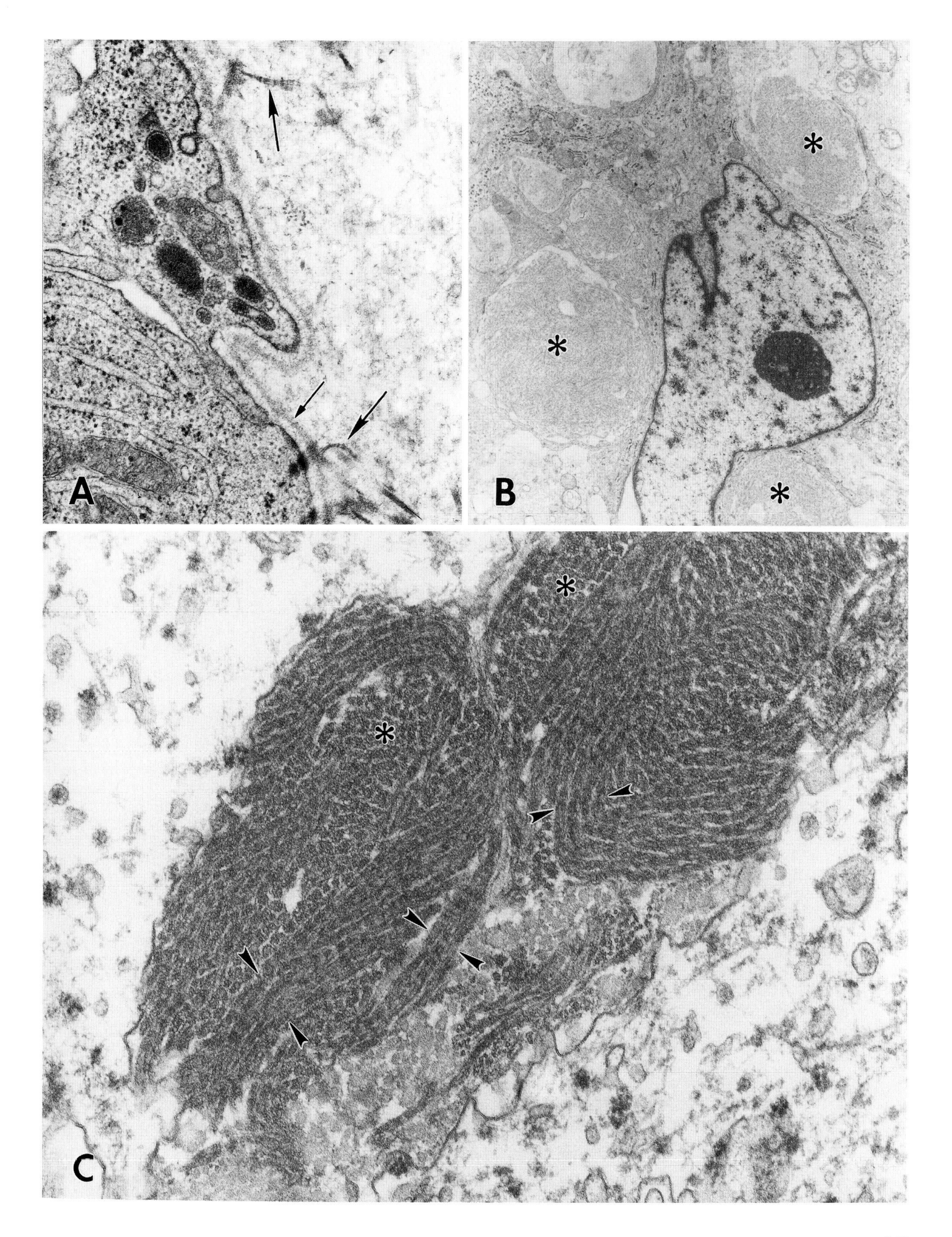
A
B
C

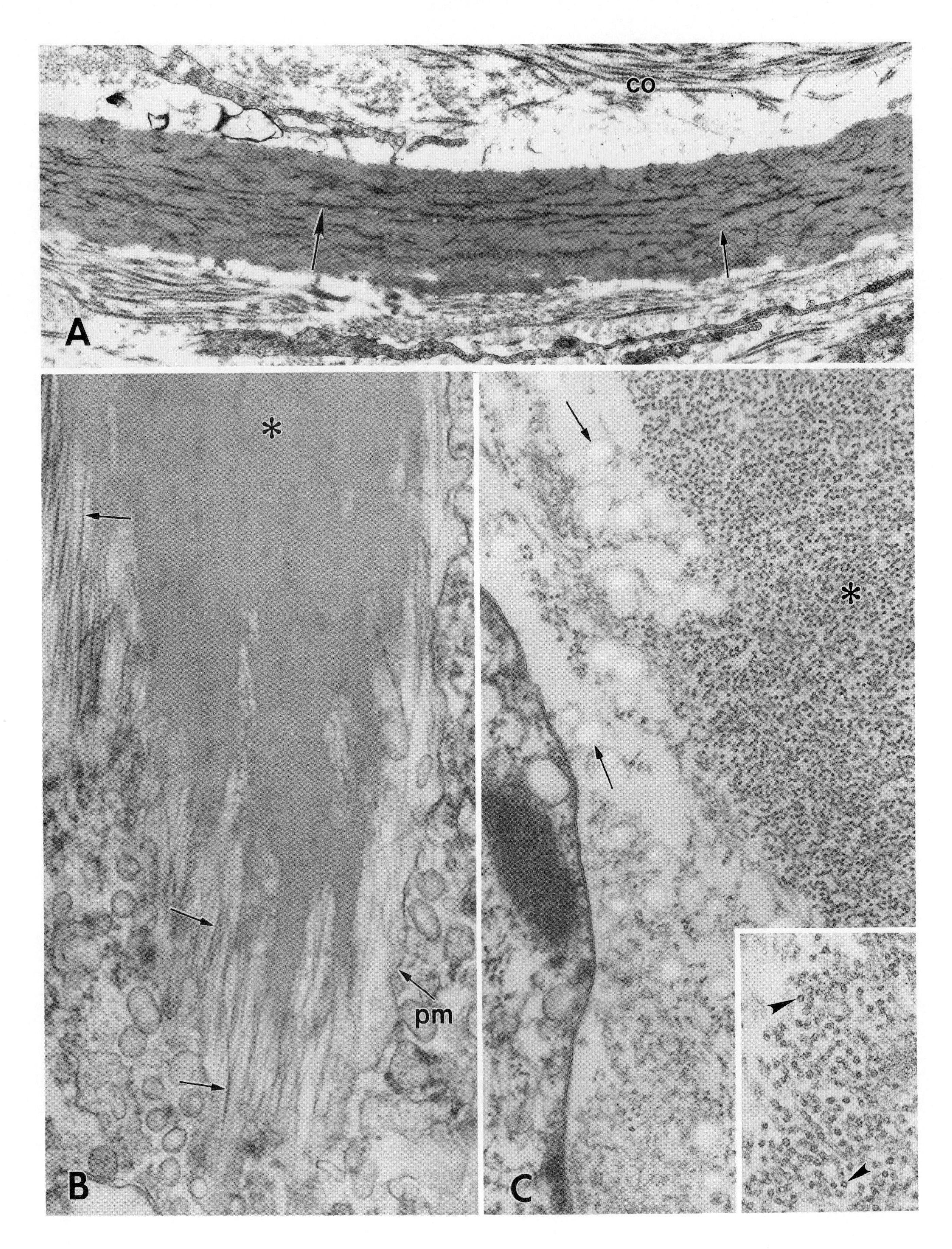
co
A
*
pm
B
*
C

present in less conspicuous quantities in a wide range of soft-tissue tumors.

After conventional processing, proteoglycans undergo varying degrees of artifactual collapse, forming granulofilamentous structures (Plate 79C and 79D), either free in the ECM or associated with matrix fibrils such as collagen.

Plate 77. **A**: An elastic fiber containing amorphous moderately dense elastin (*small arrow*) and a dense fibrillar component (*large arrow*). Note surrounding collagen. Dermis, close to epithelioid sarcoma in scalp. ×13,300. [Reproduced from Dardick I. *Handbook of Diagnostic Electron Microscopy for Pathologists-in-Training*, New York, Igaku-Shoin, 1996, with permission.] **B**: Elastic fiber showing finely textured elastin (*) and peripheral elastin-associated filaments (*arrows*). Note plasmalemma of nearby cell. Normal connective tissue of vulva. ×60,800. **C**: Part of a large oxytalan fibril (*) in cross section, consisting of orientated oxytalan filaments. Between this oxytalan fibril and the surface of the nearby cell are looser aggregates of oxytalan filaments mingling with pale-staining collagen (*arrows*). Malignant schwannoma, thigh. ×60,800. **Inset**: Note the hollow nature of the oxytalan filaments (*arrowheads*). ×93,800. co, collagen; pm, plasma membrane.

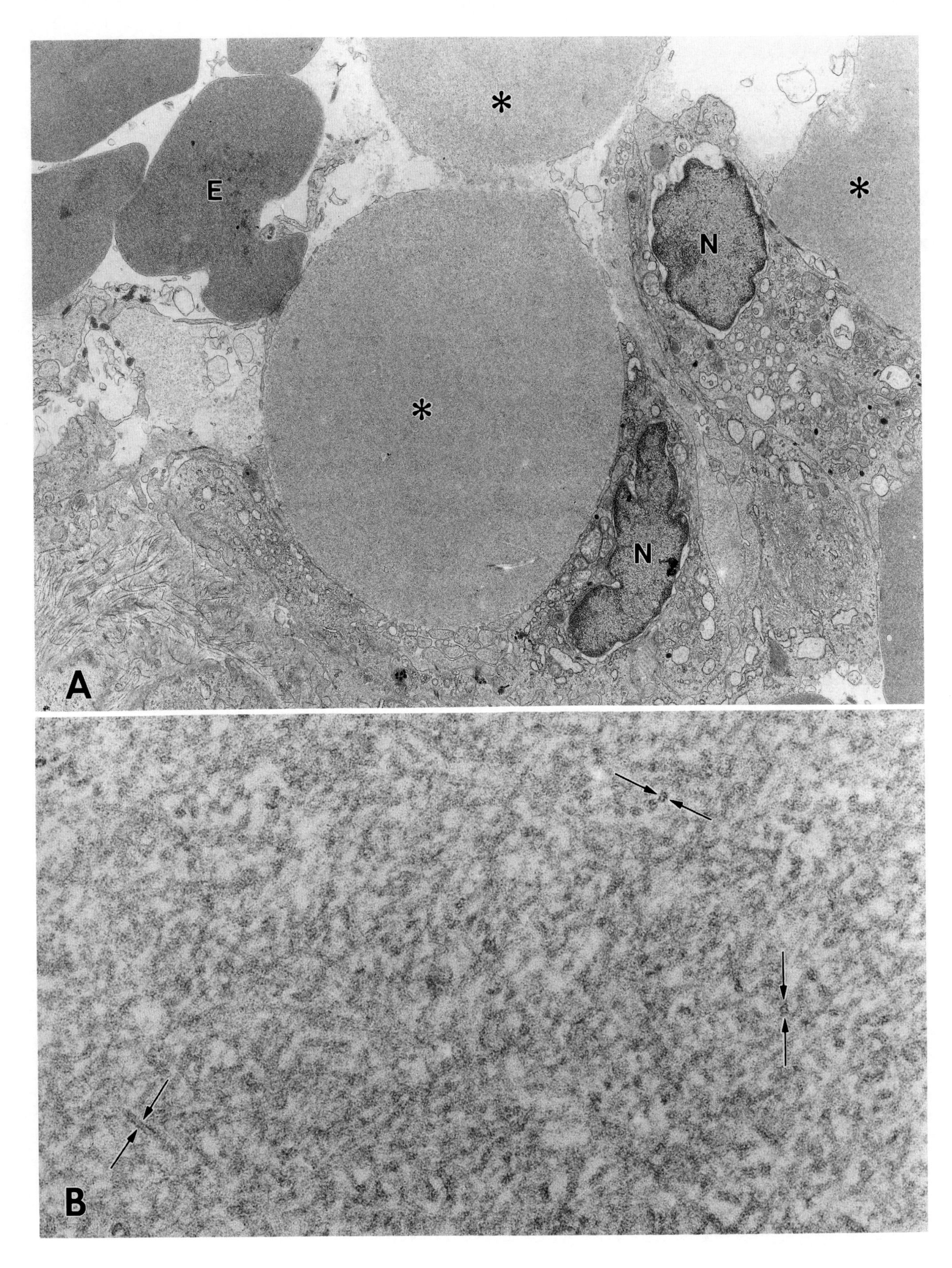
*
E
*
*
N
N
A
B

Plate 78. **A**: Homogeneous, fairly light-staining, large rounded aggregates of amyloid (*). Inflammatory cells (with nuclei) and erythrocytes are also present. Dermal nodule of amyloid, skin of face. × 7000. **B**: High magnification view to show hollow nature to amyloid fibrils (*arrows*). In this case, the amyloid fibrils are rather short. Lymphoplasmacytoid lymphoma, axillary node. × 83,300. E, erythrocyte; N, nucleus.

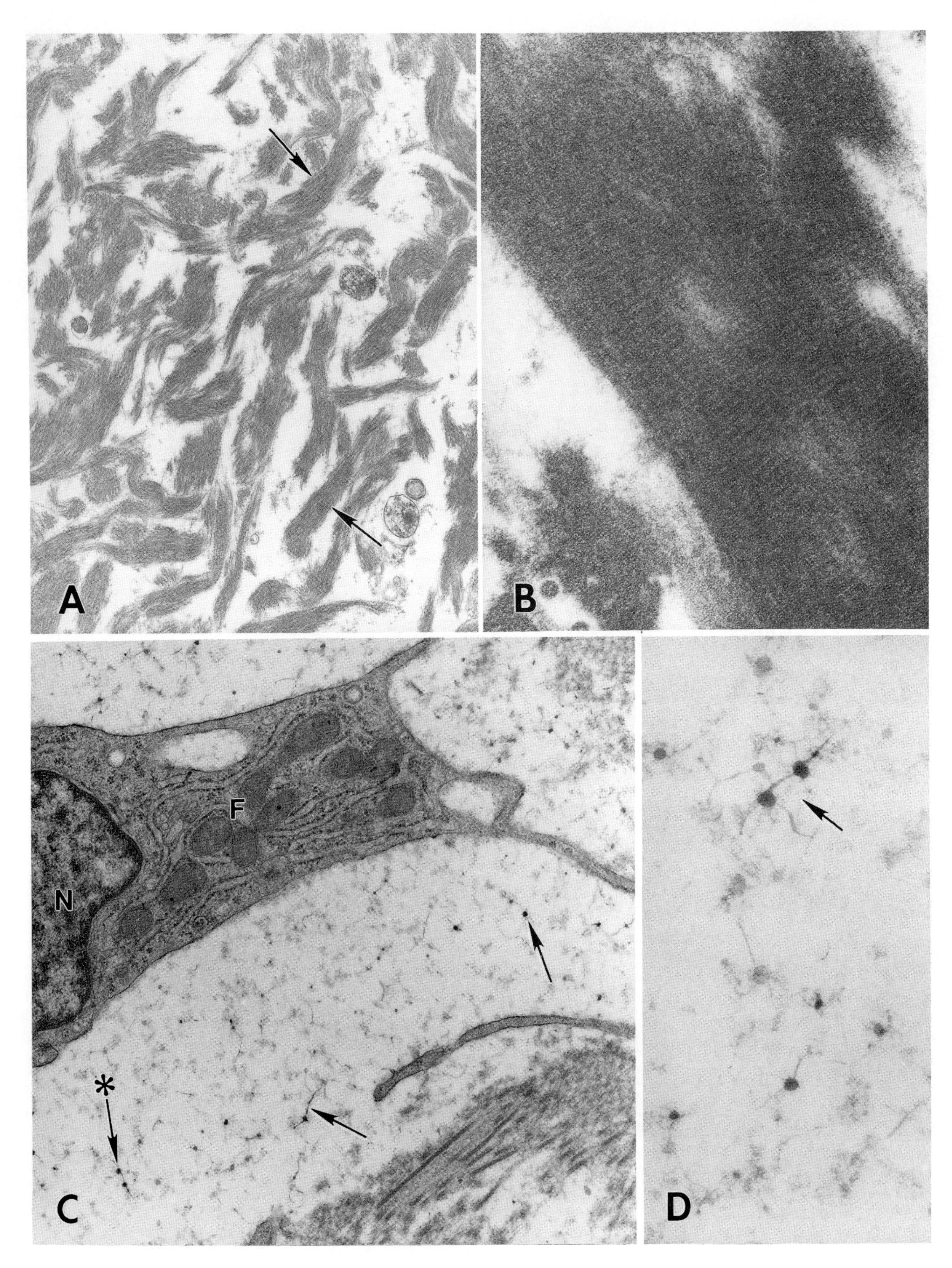
A
B
F
N
C
D

Plate 79. **A**: Fibrin at low power to show overall fibril-like appearance (*arrows*). Epithelioid hemangioma, tibia. ×16,000. **B**: Fibrin at high magnification revealing cross-striations. Gastrointestinal autonomic nerve tumor, retroperitoneum. ×95,300. **C**: Fibroblast and surrounding matrix containing proteoglycans (*arrows*). Normal breast. ×20,100. **D**: Detail of asterisked area from **C** showing granulofilamentous proteoglycan. ×69,000. F, fibroblast; N, nucleus.

References

Chapter 1: Introductory Remarks on Diagnosing Tumors by Electron Microscopy

Chan JKC, Banks PM, Cleary ML, et al: A proposal for classification of lymphoid neoplasms (by the International Lymphoma Study Group). *Histopathology* 25:517–536, 1995.

Chapter 2: Nucleus

Dickersin GR, Vickery AL, Smith SB: Papillary carcinoma of the thyroid, oxyphil type, "clear cell" variant. *Am J Surg Pathol* 4:501–509, 1980.

Duquene L, Dourov N: Ultrastructural study of nuclear bodies of the rat parotid glands. *J Submicrosc Cytol Pathol* 23:175–183, 1991.

Elavethil LJ, Kahn HJ, Hanna W: Primary multilobulated B-cell lymphoma of the breast. *Arch Pathol Lab Med* 113:1081–1084, 1989.

Ioachim HL: Non-Hodgkin's lymphoma in pregnancy. *Arch Pathol Lab Med* 109:803–809, 1985.

Tykocinski M, Schinella R, Greco MA: The pleomorphic cells of advanced mycosis fungoides. *Arch Pathol Lab Med* 108:387–391, 1984.

Ward JM, Kulwich BA, Reznik G, et al: Malignant fibrous histiocytoma. An unusual neoplasm of soft-tissue origin in the rat that is different from the human counterpart. *Arch Pathol Lab Med* 105:313–318, 1981.

Yunis EJ, Agostini RM, Devine WA: Studies on the nature of fibrillar nuclei. *Am J Pathol* 115:84–91, 1984.

Chapter 3: Endoplasmic Reticulum

Bockus D, Remington F, Luu J, et al: Induction of cylindrical confronting cisternae (AIDS inclusions) in Daudi lymphoblastoid cells by recombinant α-interferon. *Hum Pathol* 19:78–82, 1988.

Callea F, De Vos R, Togni R, et al: Fibrinogen inclusions in liver cells: a new type of ground-glass hepatocytes. Immune light and electron microscopic characterization. *Histopathology* 10:65–73, 1986.

Carstens PHB, Alexander RW, Erlandson RA, et al: Honeycomb structures in tumor cells. *Ultrastruct Pathol* 6:99–103, 1984.

Chandra S: Undulating tubules associated with endoplasmic reticulum in pathologic tissues. *Lab Invest* 18:422–428, 1968.

Cooney TP, Hwang WS, Robertson DI, et al: Monophasic synovial sarcoma, epithelioid sarcoma and chordoid sarcoma: ultrastructural evidence for a common histogenesis, despite light microscope diversity. *Histopathology* 6:163–190, 1982.

Cornog JL, Gonatus NK: Ultrastructure of rhabdomyoma. *J Ultrastruct Res* 20:433–450, 1967.

Djaldetti M: Hairy cell leukemia. *Case Histories in Human Medicine*, no. 9. Eindhoven, The Netherlands: Philips Electron Optics, 1976.

Ellis HA, Coaker T: Ultrastructure of parathyroid autografts in chronic renal failure including the occurrence of concentric membranous bodies and intermediate filaments. *Histopathology* 14: 401–407, 1989.

Eyden BP, Cross, PA, Harris M: The ultrastructure of signet-ring cell non-Hodgkin's lymphoma. *Virchows Arch A Pathol Anat* 417:395–404, 1990.

Eyden B, Prescott R, Curry A, et al: Unusual organelles in an epithelioid angiosarcoma. *Ultrastruct Pathol* 17:153–159, 1993.

Favre L, Jacot-des-Combes E, Morel P, et al: Primary aldosteronism with bilateral adrenal adenomas. *Virchows Arch A Pathol Anat Histol* 388:229–236, 1980.

Gaffney EF, Majmudar B, Hertzler GL, et al: Ovarian granulosa cell tumors—immunohistochemical localization of estradiol and ultrastructure, with functional correlations. *Obstet Gynecol* 61:311–321, 1983.

Ghadially FN, Senoo A, Fuse Y, et al: A serial section study of tubular confronting cisternae (so-called "test-tube and ring-shaped forms") in AIDS. *J Submicrosc Cytol Pathol* 19:175–183, 1987.

Gompel C: Ultrastructure of endometrial carcinoma. Review of fourteen cases. *Cancer* 28:745–754, 1971.

Graf R, Heitz PU: Tubuloampullar structures associated with the endoplasmic reticulum in pancreatic B-cells. *Cell Tissue Res* 208:507–510, 1980.

Grimley PM, Schaff Z: Significance of tubuloreticular inclusions in the pathobiology of human diseases. *Pathobiol Ann* 6:221–257, 1976.

Gyorkey F, Sinkovics JG, Gyorkey P: Electron microscopic observations on structures resembling myxovirus in human sarcomas. *Cancer* 27:1449–1454, 1971.

Hearn SA, Kontozoglou TE: Postembedding immunogold electron microscopy for S100, tubulin, and cytoskeletal proteins in an amelanotic malignant melanoma. *Mod Pathol* 2:46–55, 1989.

Kostianovsky M, Ghadially FN: Intracisternal tubules in a case of chronic lymphocytic leukemia. *J Submicrosc Cytol* 19:509–514, 1987.

Kostianovsky M, Orenstein JM, Schaff Z, et al: Cytomembranous inclusions observed in acquired immunodeficiency syndrome. Clinical and experimental review. *Arch Pathol Lab Med* 111:218–223, 1987.

Matsuda M, Nagashima K: Cytoplasmic tubular inclusion in ganglioneuroma. *Acta Neuropathol (Berl)* 64:81–84, 1984.

Nakamura S, Suzuki R, Asai J, et al: Observations on the fine structure of interdigitating cell sarcoma. *Virchows Arch A Pathol Anat* 414: 121–128, 1989.

Nakanishi I, Katsuda S, Okada Y, et al: The presence of intracytoplasmic confronting cisternae of the endoplasmic reticulum in osteosarcoma cells in interphase. *Acta Pathol Jpn* 36:261–267, 1986.

Ohno T, Park P, Higaki S, et al: Smooth tubular aggregates associated with plasmalemmal invagination in alveolar soft part sarcoma. *Ultrastruct Pathol* 18:383–388, 1994.

Schlosnagle DC, Kratochvil FJ, Weathers DR, et al: Intraoral multifocal adult rhabdomyoma. *Arch Pathol Lab Med* 107:638–642, 1983.

Seman G: Ultrastructural study of B-bodies in leukapheresed cells of patients with acute leukemia. *Oncology* 38: 204–209, 1981.

Sohval AR, Churg J, Gabrilove JL, et al: Ultrastructure of feminizing testicular Leydig cell tumors. *Ultrastruct Pathol* 3:335–345, 1982.

Suzuki T, Kaneko H, Kojima K, et al: Extraskeletal myxoid chondrosarcoma characterized by microtubular aggregates in the rough endoplasmic reticulum and tubulin immunoreactivity. *J Pathol* 156:51–57, 1988.

Szakacs JE, Kasnic G, Walling AK: Soft tissue sarcoma with complex membranous and microtubular inclusions. *Ann Clin Lab Sci* 21:430–440, 1991.

Tokue A, Yonese Y, Mato M, et al: Unusual intracytoplasmic lamellar bodies in a malignant gonadel stromal tumor. *Virchows Arch [Cell Pathol]* 49:261–267, 1985.

Uzman BG, Saito H, Kasac M: Tubular arrays in the endoplasmic riticulum in human tumor cells. *Lab Invest* 24:492–498, 1971.

Chapter 4: Golgi Apparatus

Kojimahara M, Kamita Y: Endothelial cells in the rectal venous plexus. An electron microscopic study. *J Submicrosc Cytol* 18:807–814, 1986.

Chapter 5: Neuroendocrine Granules

Bertoni-Freddari C, Fattoretti P, Pieroni M, et al: Enlargement of synaptic size as a compensative reaction in aging and dementia. *Pathol Res Pract* 188:612–615, 1992.

Ojeda VJ, Spagnolo DV, Vaughan RJ: Palisades in primary cerebral neuroblastoma simulating so-called polar spongioblastoma. A light and electron microscopical study of an adult case. *Am J Surg Pathol* 11:316–322, 1987.

Payne CM: Use of the uranaffin reaction in the identification of neuroendocrine granules. *Ultrastruct Pathol* 17:49–82, 1993.

Vassallo G, Capella C, Solcia E: Grimelius' silver stain for endocrine cell granules, as shown by electron microscopy. *Stain Technol* 46:7–13, 1971.

Chapter 6: Primary Lysosomes

Schmidt U, Mlynek M-L, Leder L-D: Electron-microscopic characterization of mixed granulated (hybridoid) leucocytes of chronic myeloid leukaemia. *Br J Haematol* 68:175–180, 1988.

Chapter 7: Mucigen Granules

Dvorak AM, Dickersin GR: Crohn's disease: transmission electron microscopic studies. I. Barrier function. Possible changes related to alterations of cell coat, mucous coat, epithelial cells, and Paneth cells. *Hum Pathol* 11:561–571, 1980.

Chapter 8: Serous/Zymogen Granules

Klimstra DS, Rosai J, Heffess CS: Mixed acinar–endocrine carcinomas of the pancreas. *Am J Surg Pathol* 18:765–778, 1994.

Chapter 9: Miscellaneous Granules

Abramowsky CR, Cebelin M, Choudhury A, et al: Undifferentiated (embryonal) sarcoma of the liver with alpha-1-antitrypsin deposits: immunohistochemical and ultrastructural studies. *Cancer* 45:3108–3113, 1980.

Dickersin GR, Oliva E, Young RH: Endometrioid-like variant of ovarian yolk sac tumor with foci of carcinoid: an ultrastructural study. *Ultrastruct Pathol* 19:421–429, 1995.

Nakanishi I, Kawahara E, Kajikawa K, et al: Hyaline globules in yolk sac tumor. Histochemical, immunohistochemical and electron microscopic studies. *Acta Pathol Jpn* 32:733–739, 1982.

Scroggs MW, Roggli VL, Fraire AE, et al: Eosinophilic intracytoplasmic globules in pulmonary adenocarcinomas: a histochemical, immunohistochemical, and ultrastructural study of six cases. *Hum Pathol* 20:845–849, 1989.

Chapter 10: Melanosomes

Bhutta S: Electron microscopy in the evaluation of melanocytic tumours. *Semin Diagn Pathol* 10:92–101, 1993.

Erlandson RA: Ultrastructural diagnosis of amelanotic malignant melanoma: aberrant melanosomes, myelin figures or lysosomes? *Ultrastruct Pathol* 11:191–208, 1987.

Eyden BP: Collagen secretion granules in reactive stromal myofibroblasts, with preliminary observations on their occurrence in spindle cell tumours. *Virchows Arch A Pathol Anat* 415:437–445, 1989.

Hunter JAA, Paterson WD, Fairley DJ: Human malignant melanoma. Melanosomal polymorphism and the ultrastructural DOPA reaction. *Br J Dermatol* 98:381–390, 1978.

Llombart-Bosch A, Carda C, Peydro-Olaya A, et al: Pigmented esthesioneuroblastoma showing dual differentiation following transplantation in nude mice. An immunohistochemical, electron microscopical, and cytogenetic analysis. *Virchows Arch A Pathol Anat* 414:199–208, 1988.

Mackay B, Ayala AG: Intracisternal tubules in human melanoma cells. *Ultrastruct Pathol* 1:1–6, 1980.

Nakatani T, Beitner H: A transmission electron microscopic study of macromelanosomes in sporadic dysplastic naevi. *Okajimas Folia Anat Jpn* 69:109–114, 1992.

Ortega VV, Diaz FM, Romero CC, et al: Abnormal melanosomes: ultrastructural markers of melanocytic atypia. Ultrastruct Pathol 19:119–128, 1995.

Schneider BV, Schnyder UW: The ultrastructure of congenital naevocytic naevi. III. Morphological variability of melanosomes. *Arch Dermatol Res* 283:438–444, 1991.

Sondergaard K, Henschel A, Hou-Jensen K: Metastatic melanoma with balloon cell changes: an electron microscopic study. *Ultrastruct Pathol* 1:357–360, 1980.

Chapter 11: Weibel–Palade Bodies

Eyden BP: Collagen secretion granules in reactive stromal myofibroblasts, with preliminary observations on their occurrence in spindle cell tumours. *Virchows Arch A Pathol Anat* 415:437–445, 1989.

Eyden BP: Ultrastructural observations on Weibel–Palade bodies suggesting exocytosis. *J Submicrosc Cytol Pathol* 25:145–148, 1993.

Fujioka Y, Tateyama M, Fujita M, et al: Weibel–Palade body-like lamellar structure in angiosarcoma. *Ultrastruct Pathol* 19:137–143, 1995.

Kitagawa M, Matsubara O, Song S-Y, et al: Neoplastic angioendotheliosis. Immunohistochemical and electron microscopic findings in three cases. *Cancer* 56:1134–143, 1985.

Chapter 12: Collagen Secretion Granules; Intracellular Collagen

Dini G, Grappone C, Del Rosso M, et al: Intracellular collagen in fibroblasts of Peyronie's disease. *J Submicrosc Cytol* 18:605–611, 1986.

Eyden BP: Collagen secretion granules in reactive stromal myofibroblasts, with preliminary observations on their occurrence in spindle cell tumours. *Virchows Arch A Pathol Anat* 415:437–445, 1989.

Eyden BP, Banerjee SS, Harris M, et al: A study of spindle cell sarcomas showing myofibroblastic differentiation. *Ultrastruct Pathol* 15:365–376, 1991.

Chapter 13: Endocytosis and Secondary Lysosomes

Takeya M, Takahashi K: Emperipolesis in a case of malignant lymphoma: electron microscopic and immunohistochemical investigation. *Ultrastruct Pathol* 12:651–658, 1988.

Chapter 14: Langerhans Cell (Birbeck) Granules

Dardick I, Dardick A, Robb I, et al: Role of electron microscopy in head and neck pathology in adults and children. *Ultrastruct Pathol* 17:427–442, 1993.

El-Labban NG: The nature of Langerhans cell granules: an ultrastructural study. *Histopathology* 6:317–325, 1982.

Chapter 15: Multivesicular Bodies

Cross PA, Eyden BP, Harris M: Signet ring cell lymphoma of T cell type. *J. Clin Pathol* 42:239–245, 1989.
Eyden BP, Cross PA, Harris M: The ultrastructure of signet-ring cell non-Hodgkin's lymphoma. *Virchows Arch A Pathol Anat* 417:395–404, 1990.
Le Tourneau A, Audouin J, Diebold J, et al.: Large anaplastic cell Ki-1 positive malignant lymphoma with peculiar endocytotic vacuoles. *Pathol Res Pract* 186:784–792, 1990.

Chapter 16: Vacuoles

Bonsib SM, Bray C, Timmerman TG: Renal chromophobe cell carcinoma: limitations of paraffin-embedded tissue. *Ultrastruct Pathol* 17:529–536, 1993.
Ishimaru Y, Fukuda S, Kurano R, et al: Follicular thyroid carcinoma with clear cell change showing unusual ultrastructural features. *Am J Surg Pathol* 12:240–246, 1988.
Yang H-Y, Wasielewski JF, Lee W, et al: Angiosarcoma of the heart: ultrastructural study. *Cancer* 47:72–80, 1981.

Chapter 17: Secondary Lysosomes

August C, Holzhausen H-J, Schroder S: Renal parenchymal malakoplakia: ultrastructural findings in different stages of morphogenesis. *Ultrastruct Pathol* 18:483–491, 1994.
Dvorak AM, Dickersin GR: Crohn's disease: transmission electron microscopic studies. I. Barrier function. Possible changes related to alterations of cell coat, mucous coat, epithelial cells, and Paneth cells. *Hum Pathol* 11(suppl):561–571, 1980.
Eyden BP, Richmond I, Hale R, et al: Lipid-rich residual bodies in human myometrium: qualitative observations. *J Submicrosc Cytol Pathol* 23:585–594, 1991.
Lee RE, Ellis LD: The storage cells of chronic myelogenous leukemia. *Lab Invest* 24:261–264, 1971.
McDuffie NG: Crystalline patterns in Auer bodies and specific granules of human leukocytes. *J Microsc (Paris)* 6:321–330, 1967.
Mintz U, Djaldetti M, Rozenszajn L, et al: Giant lysosome-like structures in promyelocytic leukemia. Ultrastructural and cytochemical observations. *Biomedicine* 19:426–430, 1973.
Mooi WJ, Dingemans KP, van den Bergh Weerman MA, et al: Ultrastructure of the liver in the cerebrohepatorenal syndrome of Zellweger. *Ultrastruct Pathol* 5:135–144, 1983.
Stavem P, Hovig T, Rorvik TO: Inclusions in bone marrow cells. *Ultrastruct Pathol* 2:389–393, 1981.
Wills EJ, Kirwan PD, Brammah S: Electron microscopy what izzits revisited: an Ultrapath VI quiz. *Ultrastruct Pathol* 18:301–325, 1994.
Yamazaki K, Eyden BP: Lipid-rich residual bodies in the human thyroid gland: ultrastructural, histochemical, and morphometric study. *J Submicrosc Cytol Pathol* 26:553–563, 1994.

Chapter 18: Mitochondria

Cross PA, Eyden BP, Harris M: Signet ring cell lymphoma of T cell type. *J Clin Pathol* 42: 239–245, 1989.
Dardick I, Claude A, Parks WR, et al: Warthin's tumor: an ultrastructural and immunohistochemical study of basilar epithelium. *Ultrastruct Pathol* 12:419–432, 1988.
Davy CL, Dardick I, Hammond E, et al: Relationship of clear cell oncocytoma to mitochondrial-rich (typical) oncocytomas of parotid salivary gland. An ultrastructural study. *Oral Surg Oral Med Oral Pathol* 77:469–478, 1994.
Kay S, Armstrong KS: Oncocytic tubular adenoma of the kidney—report of three cases with 28-year follow-up on one. *Prog Surg Pathol* 11:259–268, 1980.
Schochet SS, Lampert PW: Diagnostic electron microscopy of skeletal muscle. In: Trump BF, Jones RT, eds. *Diagnostic Electron Microscopy*, vol. 1. New York: John Wiley & Sons, 1978, pp. 238–240.
Variend S, Gerrard M, Norris PD, et al: Intra-abdominal neuroectodermal tumour of childhood with divergent differentiation. *Histopathology* 18:45–51, 1991.

Chapter 19: Lipid

El-Labban NG, Wood RD: On the so-called membrane coating granules in keratinized lichen planus lesions of the buccal mucosa. *Histopathology* 6:717—731, 1982.
Elleder M: Niemann-Pick disease. *Pathol Res Pract* 185:293–328, 1989.
Fehrenbach H, Ochs M, Richter J: Energy-filtering TEM in the fine-structural study of the mammalian lung. *Microsc Anal* September:11–14, 1995.
Ferey L, Herin P, Marnay J, et al: Histology and ultrastructure of the human esophageal epithelium. I. Normal and parakeratotic epithelium. *J Submicrosc Cytol* 17:651–665, 1985.
Ghadially FN, Harawi S, Khan W: Diagnostic ultrastructural markers in alveolar cell carcinoma. *J Submicrosc Cytol* 17:269–278, 1985.
Glauert AM: *Practical Methods in Electron Microscopy*, vol. 3, part 1: *Fixation, Dehydration and Embedding of Biological Specimens*. Amsterdam: North-Holland, 1974.
Hull MT, Eble JN: Myelinoid lamellated cytoplasmic inclusions in human renal adenocarcinomas: an ultrastructural study. *Ultrastruct Pathol* 12:41–48, 1988.
Kishikawa T: Phospholipase activities of surfactant fractions and their role in the morphological change in surfactants *in vitro*. *J Submicrosc Cytol Pathol* 22:507–513, 1990.
Martin J-J, Ceuterick Ch: The contribution of pathology to the study of storage disorders. *Pathol Res Pract* 183:375–385, 1988.

Noguchi M, Kodama T, Shimosato Y, et al: Papillary adenoma of type 2 pneumocytes. *Am J Surg Pathol* 10:134–139, 1986.

Resibois A, Tondeur M, Mockel S, et al: Lysosomes and storage diseases. *Int Rev Exp Pathol* 9:93–149, 1970.

Tome FMS, Fardeau M, Lenoir G: Ultrastructure of muscle and sensory nerve in Fabry's disease. *Acta Neuropathol (Berl)* 38:187–194, 1977.

Williams M: Conversion of lamellar body membranes into tubular myelin in alveoli of rat lungs. *J Cell Biol* 72:260–277, 1977.

Chapter 20: Glycogen

Wills EJ: Ground glasslike hepatocytes produced by glycogen-membrane complexes ("glycogen bodies"). *Ultrastruct Pathol* 16:491–503, 1992.

Chapter 21: Peroxisomes

Caballero T, Aneiros J, Lopez-Caballero J, et al: Fibrolamellar hepatocellular carcinoma. An immunohistochemical and ultrastructural study. *Histopathology* 9:445–456, 1985.

De Craemer D, Pauwels M, Hautekeete M, et al: Alterations of hepatocellular peroxisomes in patients with cancer. Catalase cytochemistry and morphometry. *Cancer* 71:3851–3858, 1993.

Chapter 22: Contractile Filaments

Agamanolis DP, Dasu S, Krill CE: Tumors of skeletal muscle. *Hum Pathol* 17:778–795, 1986.

Cross PA, Eyden BP, Joglekar VM: Carcinosarcoma of the urinary bladder. A light, immunohistochemical and electron microscopical case report. *Virchows Arch A Pathol Anat* 415:91–95, 1989.

Eyden BP, Harris M, Banerjee SS, et al: The ultrastructure of epithelioid sarcoma. *J Submicrosc Cytol Pathol* 21: 281–293, 1989.

Eyden BP, Hale RJ, Richmond I, et al: Cytoskeletal filaments in the smooth muscle cells of uterine leiomyomata and myometrium: an ultrastructural and immunohistochemical analysis. *Virchows Arch A Pathol Anat* 420:51–58, 1992.

Kuwashima Y, Hayashi S, Arata M, et al: Rhabdomysosarcoma with focal cartilaginous differentiation (malignant mesenchymoma) of the inferior vena cava. *Acta Pathol Jpn* 42:382–385, 1992.

Chapter 23: Nemaline Rods; Leptomeric Fibrils

Fawcett DW: The sporadic occurrence in cardiac muscle of anomalous Z bands exhibiting a periodic structure suggestive of tropomyosin. *J Cell Biol* 36:266–270, 1968.

Chapter 24: Intermediate Filaments

Harris M, Eyden BP, Joglekar VM: Rhabdoid tumour of the bladder: a histological, ultrastructural and immunohistochemical study. *Histopathology* 11:1083–1092, 1987.

Lowe J, Blanchard A, Morrell K, et al: Ubiquitin is a common factor in intermediate filament inclusion bodies of diverse type in man, including those of Parkinson's disease, Pick's disease, and Alzheimer's disease, as well as Rosenthal fibres in cerebellar astrocytomas, cytoplasmic bodies in muscle, and Mallory bodies in alcholic liver disease. *J Pathol* 155:9–15, 1988.

Chapter 25: Microtubules

El-Labban NG, Rindum J, Nielsen H, et al: Crystalline inclusions in epithelial cells of hairy leukoplakia: a new ultrastructural finding. *Scand J Dent Res* 96:353–359, 1988.

Gonzalez S, von Bassewitz DB, Grundmann E, et al: Rudimentary cilia in hyperplastic, metaplastic and neoplastic cells of the lung and pleura. *Pathol Res Pract* 180:511–515, 1985.

Lurie M, Rennert G, Goldenberg S, et al: Ciliary ultrastructure in primary ciliary dyskinesia and other chronic respiratory conditions: the relevance of microtubular abnormalities. *Ultrastruct Pathol* 16:547–553, 1992.

Mierau GW, Agostini R, Beals TF, et al: The role of electron microscopy in evaluating ciliary dysfunction: report of a workshop. *Ultrastruct Pathol* 16:245–254, 1992.

Pearl GS, Takei Y, Kurisaka M, et al: Cystic prolactinoma. A variant of "transitional cell tumor" of the pituitary. *Am J Surg Pathol* 5:85–90, 1981.

Chapter 26: Intercellular and Cell-to-Matrix Junctions

Burns TR, Johnson EH, Cartwright J, et al: Desmosomes of epithelial malignant mesothelioma. *Ultrastruct Pathol* 12:385–388, 1988.

Garfield RE, Hayashi RH: Appearance of gap junctions in the myometrium of women during labor. *Am J Obstet Gynecol* 140:254–260, 1981.

McNutt NS: Ultrastructure of intercellular junctions in adult and developing cardiac muscle. *Am J Cardiol* 25:169–183, 1970.

Moll R, Cowin P, Kapprell H-P, et al: Desmonsomal proteins: new markers for identification and classification of tumors. *Lab Invest* 54:4–25, 1986.

Pitt MA, Wells S, Eyden BP: Carcinosarcoma arising in a duct papilloma. *Histopathology* 26:81–84, 1995.

Quinonez G, Simon GT: Cellular junctions in a spectrum of human malignant neoplasms. *Ultrastruct Pathol* 12:389–405, 1988.

Chapter 27: Lamina: Fibronexus

Carnay JA: Psammomatous melanotic schwannoma. A distinctive heritable tumor with special associations, including cardiac myxoma and the Cushing syndrome. *Am J Surg Pathol* 14:206–222, 1990.

Damjanov I, Tuma B, Dominis M: Electron microscopy of large cell undifferentiated and giant cell tumors. *Pathol Res Pract* 184:137–160, 1989.

Dickersin GR: The electron microscopic spectrum of nerve sheath tumors. *Ultrastruct Pathol* 11:103–146, 1987.

Eyden BP: A brief review of the fibronexus and its significance for myofibroblastic differentiation and tumor diagnosis. *Ultrastruct Pathol* 17:613–624, 1993.

Papadimitriou JC, Drachenberg CB: Ultrastructural features of the matrix of small cell osteosarcoma. *Hum Pathol* 25:430, 1994.

Reale E: Electron microscopy of the basement membranes. In: Ruggeri A, Motta PM, eds. *Ultrastructure of the Connective Tissue Matrix.* Boston; Martinus Nijhoff, 1984, pp. 192–211.

Stewart KR, Casey MJ, Gondos B: Endodermal sinus tumor of the ovary with virilization. Light- and electron-microscopic study. *Am J Surg Pathol* 5:385–391, 1981.

Chapter 28: Processes: Lumina

Borg-Grech A, Morris JA, Eyden BP: Malignant osteoclastoma-like giant cell tumor of the renal pelvis. *Histopathology* 11:415–425, 1987.

Erlandson RA, Huvos AG: Chondrosarcoma: a light and electron microscopic study. *Cancer* 34:1642–1652, 1974.

Henderson DW, Shilkin KB, Whitaker D, et al: The pathology of malignant mesothelioma, including immunohistology and ultrastructure. In: Henderson DW, Shilkin KB, Langlois SLeP, et al; eds. *Malignant Mesothelioma.* New York: Hemisphere, 1992, pp. 69–139.

Storkel S, Steart PV, Drenckhahn D, et al: The human chromophobe cell renal carcinoma: its probable relation to intercalated cells of the collecting duct. *Virchows Arch B Cell Pathol* 56:237–245, 1989.

Waston RJ, Eyden BP, Howell A, et al: Ultrastructural observations on the basal lamina in the normal human breast. *J Anat* 156:1–10, 1988.

Chapter 29: Crystals

Carson HJ, Buschmann RJ, Weisz-Carrington P, et al: Identification of Charcot–Leyden crystals by electron microscopy. *Ultrastruct Pathol* 16:403–411, 1992.

Eyden BP: Critical commentary: Fibril formation in the rER of lymphoma cells. *Pathol Res Pract* 190:90–94, 1994.

Jay V, Edwards V, Rutka JT: Crystalline inclusions in a subependymal giant cell tumor in a patient with tuberous sclerosis. *Ultarastruct Pathol* 17:503–513, 1993.

Kodet R: Juxtaglomerular cell tumor. An immunohistochemical, electron-microscopic, and in situ hybridization study. *Am J Surg Pathol* 18:837–842, 1994.

Mukai M, Torikata C, Iri H, et al: Crystalloids in angiomolipoma. 1. A previously unnoticed phenomenon of renal angiomolipoma occurring at a high frequency. *Am J Surg Pathol* 16:1–10, 1992.

Nistal M, Paniagua R, Abaurrea MA, et al: Hyperplasia and the immature appearance of Sertoli cells in primary testicular disorders. *Hum Pathol* 13:3–12, 1982.

Ro JY, Sahin AA, El-Naggar AK, et al: Intraluminal crystalloids in struma ovarii. Immunohistochemical, DNA flow cytometric, and ultrastructural study. *Arch Pathol Lab Med* 115:145–149, 1991.

Schaff Z, Eder G, Eder C, et al. Intracytoplasmic crystalline inclusions in the hepatocytes of humans and chimpanzees. *Ultrastruct Pathol* 14:303–309, 1990.

Schnoy N: Ultrastructure of a virilizing ovarian Leydig-cell tumor. *Virchows Arch [Pathol Anat]* 397:17–27, 1982.

Spear GS, Gubler M-C, Habib R, et al: Dark cells in cystinosis: occurrence in renal allografts. *Hum Pathol* 20:472–476, 1989.

Van Hoeven KH, Drudis T, Cranor ML, et al: Low-grade adenosquamous carcinoma of the breast. A clinicopathologic study of 32 cases with ultrstructural analysis. *Am J Surg Pathol* 17:248–258, 1993.

Chapter 30: Extracellular Matrix

Eyden BP, Ferguson J: Anchoring fibrils and type VII collagen in human breast. *J Submicrosc Cytol* 22:477–479, 1990.

Eyden BP, Harris M, Greywoode GIN, et al: Intranodal myofibroblastoma: report of a case. *Ultrastruct Pathol* 20:79–88, 1996.

Fukuda Y, Miyake H, Masuda Y, et al: Histogenesis of unique elastophilic fibers of elastofibroma: ultrastructural and immunohistochemical studies. *Hum Pathol* 18:424–429, 1987.

Hough AJ, Mottram FC, Sokoloff L: The collagenous nature of amianthoid degeneration of human costal cartilage. *Am J Pathol* 73:201–216, 1973.

Kumaratilake JS, Krishnan R, Lomax-Smith J, et al: Elastofibroma: disturbed elastic fibrillogenesis by periosteal-derived cells? An immunoelectron microscopic and in situ hybridization study. *Hum Pathol* 22:1017–1029, 1991.

Leak LV, Burke JF: Ultrastructural studies on the lymphatic anchoring filaments. *J Cell Biol* 36:129–149, 1968.

Tan SY, Pepys MB: Amyloidosis. *Histopathology* 25:403–414, 1994.

INDEX